The Israeli Air Force 1947-1960

An Illustrated History

Salvador Mafé Huertas

The Israeli Air Force 1947-1960

Salvador Mafé Huertas

Schiffer Military/Aviation History
Atglen, PA

ACKNOWLEDGMENTS

The author is in debt to the many Chel Ha'avir veterans that cooperated in the making of this book. Also, I want to express my thanks to the Chel Ha' avir; IDF Spokesman; Shlomo Klescelski; Israel Air Force Magazine; Israeli Embassy in Madrid; Imperial War Museum; RAF Historical Branch; SIRPA Air at Paris; Ben Gurion Research Center & Archives at Sde Boker, (Israel); Federico Wichi, Jose Miguel Sales, Manuel Perales, Fernando Ribes, Javier Valero, Javier Matute, Jose Sanz, Manuel Rodríguez, Rafael Trevino and Antonio Catala. Last but not least to my daughters, Mercedes and Cristina, and my Wife, Mercedes, for their encouragement and continuous support. It is not always easy to have an aviation journalist as a father or husband !

DEDICATED

TO MY FRIENDS OF THE FAMED ALAMBIQUE SQUADRON, BOTH THE FOUNDERS AND LATER ARRIVALS, AS WELL AS THE ATTACHED "DOLARES."

IN MEMORIAM

FOR *PITU III*

Cover artwork by Steve Ferguson, Colorado Springs, CO

NEW FALCONS OF DAVID: Only seven years had passed since the birth of Israel when its fledgling Air Defense force was to be tested once again in aerial combat. On August 31, 1955, two Egyptian Vampire jets flew an armed reconnaissance of southern Israel initiating an interception by two FR.9 Meteors of 117 Squadron. Within minutes, mission leader Ceren (captain) Yoeli circled in behind the Vampires and shot down the Egyptian wingman in the first firing pass, then maneuvered through several turns to finally dispatch the opposing leader. Yoeli would receive a personal *tzalash* citation from Chief of Staff Dayan for attaining the first jet kills of the IAF, foretelling of the deadly era about to erupt in the Middle East. -Research provided by Russell Mueller

Book Design by Ian Robertson.

Library of Congress Catalog Number: 97-80323

Printed in China.
ISBN: 0-7643-0390-2

We are interested in hearing from authors with book ideas on related topics.

Published by Schiffer Publishing Ltd.
4880 Lower Valley Road
Atglen, PA 19310
Phone: (610) 593-1777
FAX: (610) 593-2002
E-mail: Schifferbk@aol.com
Please write for a free catalog.
This book may be purchased from the publisher.
Please include $3.95 postage.
Try your bookstore first.

CONTENTS

Introduction:
The First Fighter Squadron Maiden Combat Mission

May 29, 1948. A mere fortnight had passed since the charismatic David Ben Gurion announced the birth of Israel, and hours later the armies of five Arab countries—plus irregulars—invaded the tiny United Nations proclaimed state. Also, eight days had passed since the first "Knife" (Avia S-199) fighter arrived at Tel Nof air base in a "Ballak's" flight Douglas C-54. The soon-to-be-named 101 Squadron, "Tayeset Ha'rishma" (the First fighter Squadron), launched the first ever combat mission late in the afternoon on that hot Saturday with the only four airworthy S-199s against a large Egyptian column advancing from the south and threatening Tel Aviv. The four pioneering fighter jocks were Lou Lenart, Modi Allon, Ezer Weizman and Eddie Cohen. The attack did not cause much damage, but it did stop the Egyptians. It had its cost; two Avias were written off, and a pilot was lost in that maiden mission, as was fully explained in the after-action report:

"At 1945 local a four aircraft formation took-off to attack a large column of Egyptian vehicles between Ishdod and Gas'ser Ishdod, which had just stopped on the southern side of a destroyed bridge:

(1) - The leader Lou Lenart approached Ishdod from the north and dropped his bombs in the center of the village. He observed the target's vehicle concentration at the road's curve some 300 meters south of Ashod. His second pass was from southeast to northwest. His third pass was from north to south and in both he strafed with his machine guns as the cannon ceased firing after the first ten rounds. Flak was very intense, most of it was from 40mm guns. Landed at 2025.

(2) - Modi Allon approached Ishdod from the northeast. South of the bridge he observed a lot of vehicles as well as east of Ashod. He dropped his bombs on the vehicles on the road. In the second pass he attacked from east to west and in his third from north to south. In these passes he finished the ammunition and returned home flying over the sea. The flak was heavy. He estimated the number of vehicles in hundreds. Upon landing at Tel Nof the left brake did not work properly and he could not maintain a straight line as a result. Finally he performed a ground loop, the left tire blew up and the wing tip struck the ground and was damaged structurally. Landed at 2005.

Ezer Weizman, actually Israel's President, became a rated pilot with the RAF late in WWII, in this picture he is seen as a Flight Sargeant. (RAF)

Mordechai Modi *Allon was 101 Squadron first Commanding Officer from May 1948 until his death on 16 October of the same year.(IAF)*

The Avia S199 was the only fighter available to the fledging Israeli Air Force for the first months of the Indepence. This example currently at the IDFAF Museum at Hatzerim air base is the only surviving machine.(Author)

(3) - Ezer Weizman attacked from south to north. At first he saw some 20 vehicles south of Ashod and dropped his first bomb. In his second pass he attacked from west to east and dropped his second bomb about one kilometer north of the town. His third pass was from south to north. His canons fired one round each and jammed. Machine guns worked fine. Landed at 2015.

(4) - Eddie Cohen was in radio contact with base. On his way back reported that all was OK, that he had the base on sight and that he was about to land. From Tel Nof he was not observed and he did not land there. Our men in Hatzor air base saw an aircraft engulfed in flames trying a crash landing two and half kilometers away. Two infantry platoons were sent immediately, but the Egyptian forces were the first to reach the location. It is thought that Eddie Cohen mistook Hatzor for Tel Nof, trying to land there with his damaged S-199."

The writing of this book, the first in-depth and fully detailed research about the early years of the Chel Ha'avir—from the light-plane days to the supersonic fighters received roughly ten years later—has been the culmination of a long research period by the author. A research made harder by the tight—but understandable and respected—security measures involving Israel's Forces. An exceptional source of information has been the help provided by many Chel Ha'Avir veterans who fought in the hard days during 1948/1949 who helped establish a viable air force in 1949-1956 in spite of tight budgets, obsolescent equipment and lack of experience; and also those who fought in Sinai in late 1956, establishing the seeds for its future role as the airborne shield of Israel.

As the reader might find, the beginnings were extremely hard, since they had to face an unfriendly attitude from both Western superpowers (the United States and the United Kingdom), which took place while many people—civilian and military—were dying in the newborn Hebrew state. But in observing that official policies do not always reflect the citizenship's feelings, most of the colorful band of foreign volunteers that helped the Chel Ha'avir–and the other defense forces–to resist, fight back and win, came from those countries. These foreign volunteers, mostly with combat experience in WWII, provided a core in which many highly talented young Israelis learned fast.

One thing was certain, then and now; Israel exists because of the resolute people that live in this small country, both civilian and military, but above all because of the Chel Ha'avir, which in the following years would be proclaimed the most combat experienced air force in the world.

Close-up of the beautifully restored S199 at the IDFAF Museum showing the position of 101 Squadron insignia.The two kill markings represent Modi *Allon´s two victories, a pair of Egyptian C-47s shot down over Tel-Aviv on 3 June 1948, although these markings were not applied at the time. (Author)*

Chapter 1:
Secret Air Force

During the evening of 14 May 1948, when David Ben Gurion, the first Prime Minister and Minister of Israel announced the birth of the new state, the Sherut Avir (Air Service)—soon to grow into the Chel Ha'avir (The Air Force)—already had some 30 light-planes, dozens of pilots and hundreds of ground crews, having also about fifteen transports (C-46 Commando, C-47 Dakota, C-54 Skymaster and C-69 constellation) abroad and ten Avia S-199 fighters on order. The Sherut Avir was the air arm of the Haggana underground movement and was thus dubbed by Ben Gurion as the "Underground Air Force."

When the United Nations decided on 29 November 1947 to end the British mandate in Palestine and to split its territory into a Jewish and an Arab state, it was quite obvious that a fierce struggle was about to take place. It was soon realized that the Jewish community would have to face not only the local Arab forces within Palestine, but also an invasion by five regular armies equipped with relatively modern weapons such as tanks, artillery, ships and aircraft—all denied from the anticipated state of Israel by the U.N. decision to embargo all arms sales to the Middle East.

The background to the Sherut Avir can be traced back to the thirties when the roots of all three branches of modern aviation—military, commercial and sporting—were laid. The sport sector was dominated by several gliding clubs, the first of which was named Ha'carmel (The Carmel, after a mountain near Haifa), and the first gliding in Palestine took place on 17 July 1932. In the mid-thirties all the remaining gliding clubs were united as Clob Ha'teoffa Le'eretz Israel (The Aero Club of Israel), to be termed from now on as the Aeroclub. The commercial branch began with a short-lived company known as Palestinian Airways, Ltd. The company was registered in the United Kingdom in December 1934, and its parent firm was Imperial Airways. Two Short S16 Scions (VQ-PAA and VQ-PAB) were the first aircraft to use the register VQ-P...and were flown by British aircrews on routes extending all over the Middle East. The Scions were operated from Lod Airport (now Ben Gurion International Airport) on regular lines to Haifa and Beirut, as well as on charter flights and local Air Days to draw attention to aviation among the local population. Since most of the Jewish community was in the Tel Aviv area (Lod was an Arabic city), Palestinian Airways obtained permission from the British Authorities to erect a small airport north of Tel Aviv, near the Reading Electricity power station, which commenced operations in 1938. This airport was actually known as Sde-Dov (Dov Field after the name of Dov Hoz, the first chairman of the Aviron Company), and was about to play a major role in the annals of Israeli aviation.

Aerial view of the Yarcom Hotel in Tel-Aviv, which was the first Chel Ha'avir HQ. (IAF)

Palestinian Airways later purchased a De Havilland DH-89 Dragon Rapide (VQ-PAC) and a Short S22 Scion Senior (VQ-PAD), but ceased operations in the summer of 1940 when Italian forces in North Africa threatened the Middle East. The British forbade civilian flying and pressed Palestinian Airways aircraft into service with the RAF's Lod Communications Flight. The three Shorts were written-off, but the Dragon Rapide survived WWII and served with the Chel Ha'avir (Israel Force/Air Force—IDFAF) until its withdrawal from use in 1955.

The third branch mentioned was the military. The importance of military aviation was quite early understood by many Jewish community leaders, especially those of the underground movements. The atmosphere was such that many isolated Jewish settlements were raided by their Arab neighbors and there was no way to send in reinforcements because of the relatively long distances be-

Men and machines from Aviron at Lod airfield. (IAF)

tween the Jewish centers and the settlements. The poor state of the roads and the fact that the Arabs controlled them and thus could establish road blocks whenever they wished was a factor. Aircraft were the obvious solution. Since 1920 the RAF had operated with the No 14 Sqn. for policing missions and the support of British ground forces from Ramle and Amman airfields. At first No 14 Sqn. was equipped with the Bristol Fighter, and from 1926 the De Havilland DH-9A. Three years later the squadron converted to the Fairey IIIF; in1932 it began operations with the Fairey Gordon, which continued until 1938 when the Vickers Wellesley was introduced. The squadron was reinforced from time to time during periods of tension between the Arabs and Jews, and especially during the Arab raisings in 1936-1939. The lonely squadron was joined at Ramle by No 6 Squadron, which operated during its duration in Palestine the Hawker Hart and Hardy, Gloster Gauntlet and Westland Lysander. No 33 Squadron joined the two others during 1938; it was equipped with Gloster Gladiator fighters and operated from Ramle and Lod. The effective use of air power by the British forces impressed the Jewish community leaders, and they thought about it as a means to project the quality edge against the numerical superiority enjoyed by the Arabs. It was soon realized that the first step towards military aviation was to train pilots, as no real cadre of Jewish aviators existed in Palestine, except for a few foreign-trained private fliers.

Simultaneously, two underground movements initiated pilot training under the cover of civilian flying schools. The small movement Irgon Tsavai Leumi (National Military Organization, better known as Atsel after its Hebrew acronym) used the company Sherutai Eretz Israel Le'teoffa (Palestinian Flying Services), which was founded in 1937 by Moshe Haim Katz. It operated a single De Havilland DH-60M Gipsy Moth (G-ABMX), flown by Katz's son Freddy, which was written-off in a landing accident at Lod on 25 June 1938. Soon afterwards, an experienced Jewish pilot, Edwin Leibovich, came to Palestine with his Taylorcraft Model A (registered VQ-PAH in December 1938), and the Atsel persuaded Leibovich and Katz to train its members.

Two Taylorcraft Model BLs were added (VQ-PAI and VQ-PAJ), and the first course included six cadets that were trained by the instructors Edwin Leibovich and Freddy Katz. The graduation ceremony was on 21 April 1939 at Lod, and only one of these six pioneers, Benjamin Cahana, served with the Chel Ha'avir; he was killed in action during the Sinai campaign of 1956. Another seven cadets gained their PPLs (Private Pilot Licenses) with Palestine Flying Services before that company ceased operations because of economical difficulties and the Taylorcrafts were sold to Aviron in August 1939. The Aviron (Aeroplane) company commenced operations in March 1938 with a single De Havilland DH-82A Tiger Moth (G-ACYN) and was owned by the Jewish Agency, which also sponsored the large Haggana movement that years later formed the basis to the Tsva Haggana Le'Israel (Israel Forces, or IDF). The Tiger Moth was flown from the U.K. by Captain Grey, a British pilot who was supposed to become the Chieng (the chief instructor) of the flying school. Fairly soon it was discovered that Captain Grey suffered from drunkenness, and he was replaced by Emmanuel Tsur, who had gained his instructor's license in France. The Tiger Moth was used to fly the local Jewish community leaders, patrol the roads, communicate with isolated settlements, guard groups of workers at remote areas and training. Aviron set up its flying school at Affikin in the Jordan Valley, but soon afterwards the sole Tiger Moth was written-off when the tired instructor Tsur fell asleep during flight. Tsur was unhurt, and three Polish RWD 8s (VQ-PAG, -PAK and SP-BLL) were obtained from the Aero Club to continue operations. Aviron expanded rapidly, and during the summer of 1939 added to its fleet the three aforementioned Taylors, two RWD 13s (VQ-PAL and VQ-PAM), and one RWD 15 (VQ-PAE). At the end of that summer, World War Two broke out.

At first there was not any influence upon flying activities at Palestine, but when Italy joined in and the Reggia Aeronautica bombed the oil refineries and harbor at Haifa, Palestine became a staging area for the British, and many air bases that today still serve the Chel Ha'avir, such as Ramat David, Tel Nof and Hatzor, were erected in those days. Civil aviation was prohibited, and many Jews joined the British service, of which 1863 went into the RAF but were restricted to groundcrew jobs such as drivers, kitchen workers, logistics, armorers and technical trades. Later on a few of them were accepted to flying courses, and some 30 gained aircrew wings, including about 15 pilots, and it was only natural that from this group three future Chel Ha'avir commanders, Aharon Remez, Dan Tolkowski and Ezer Weizman, rose. During the summer of 1941, the Ha'chalotz (The Pioneer) movement organized a civilian flying course for some twenty young American Jews that were to emigrate to Palestine. However, most of them joined the U.S. armed forces after the Japanese attack on Pearl Harbor on 7 December 1941 when civilian pilots were freed from the primary stage of the flying course. Two of the participants of that course, Heyman Shamir and Pessah Tal, were later to play key roles in the formation of the Chel Ha'avir. Both served with the USAAF, Tal as a transport and bomber pilot and Shamir as a navigator. Many other Jewish served with other air forces around the world, and quite a number of them came to help to help the birth of Israel and its air force.

In Palestine, the flying activity of the Haggana movement was reorganized, and the Palmach's (Hebrew acronym of Plogot Machatz or Victory Companies) Flying Platoon was founded in 1945 after eight Palmach members gained their PPLs with Aviron at Ramle. The Platoon was based firstly at kibbutz Mishmar Ha'emek under the command of Pinchas Ben Porat, but later on moved to kibbutz Na'an (just south of Lod airport) and the nearby RAF base at Ramle, where the platoon members were trained on the old aircraft of Aviron. During the war years a Miles M3A Falcon (VQ-PAO), a Benes Meraz Be550 Bebe (VQ-PAQ) and a Fokker F18 (VQ-PAF) were added to the Aviron's fleet, although the Fokker was not flown by Aviron. Also during WWII the Atsel bought a Zlin 12 (VQ-PAP) that was registered as a private aircraft on the name of Benjamin Cahana. As WWII ended it was apparent that a settlement to the Jewish problem was inevitable, and both the Arab and the Jewish sides had done their best to improve their opening cards. In 1945 Ha'moatcha Ha'chlalit Le'teoffa Be'eretz Israel (The Israeli General Council of Aviation) was established in order to control the flying activities that were funded by the Jewish Agency Aviron, the Aero Club and the Palmach's Flying Platoon. The council was headed by Joshua "Iseek" Ashel, a veteran Haggana member without any aeronautical background, who would later become the first commander of the Sherut Avir. The growing importance that the Haggana attached to aviation can be exemplified by the council's budget: 2500 Palestinian Pounds for 1945, 3200 for 1946 and 10000 for 1947. Moreover, when Ashel demanded 125000 PP in November 1947 for Sherut Avir's budget, he was immediately approved with 110000PP. During this period a few aircraft were added to the already-obsolete fleet of the Jewish community: an Auster J1 Autocrat (VQ-PAS), a De Havilland DH-89 Dragon Rapide (VQ-PAR), two De Havilland Canada DH-82C Tiger Moths (VQ-PAT and VQ-PAN), and a privately owned Republic RC3 Seabee (NV6731K). The RWD 13s and the Bebe were sold to private operators, and one RDW 8 (SP-BLL) was withdrawn from use in 1946. Aviron used the Rapide and the RWD 15 as air taxis, and the remaining planes for training. As the situation deteriorated, the Palmach's Flying Platoon used the old Aviron's aircraft to supply and communicate with Haggana forces in the Negev desert and with remote settlements, as well as to photograph, with hand-held cameras, potential targets for future operations. The activity had to be conducted very carefully because the British authorities disapproved of it.

Auster AOP.3 VQ-PAS from the Tel-Aviv Squadron shortly before a recce flight early in 1948. (IAF)

In the summer of 1947, Alexander Zillioni, an ex-RAF officer, sent the Haggana a memorandum in which he promoted the creation of an air arm, and less than a month later, in August, Aharon Remez and Heyman Shamir were asked to draw a plan of an air arm that could meet the threat of war. Remez was born in Israel in 1919 and served as a fighter pilot in Europe with the RAF during WWII. Shamir was born in the USSR (Kiev) in 1920 and served as a navigator with the USAAF during the war. They handed over the plan to the Haggana's staff chief, Israel Gallilee, on 23 October 1947. Remez was destined to become Chel Ha'avir commander with Shamir as his vice-commander.

This scheme was a dream in the period in which the largest aircraft owned by the Jewish community was an old Dragon Rapide. It was a dream because the estimated cost of the plan was about three million dollars in a period in which the yearly budget of aviation was 10000 PP, and it was also a dream because it was already known that the Jewish community in Palestine would have to face the strength of five regular Arab armies on the very first day of the Israeli state declaration of existence without a regular army, navy or air force, in addition to suffering from a U.N. embargo of arms.

It was a wild dream of two young friends that stood no chance of becoming a reality, but miraculously some of it turned into reality in the hands of farsighted young men that left all they had—families, jobs and futures—in order to help in the creation of the state of Israel.

An escort aircraft carrier, the USS Attu, was bought in the USA, although she never sailed. An air bridge was set up to supply the forces fighting for their lives in the newly born state, and aircraft and crews were smuggled in the face of the embargo. Much earlier, however, a small squadron of light-planes was organized, and Sherut Avir, the forerunner of the Chel Ha'avir, came into existence.

It was a time of changes; a period in which history was made. The British had decided to withdraw from Palestine and leave its future in the hands of the U.N. It was clear at that stage that the U.N. decision would be to split the land into two states—a Jewish one adjacent to an Arab one. While the Jews accepted the plan out of necessity, the Arabs opposed it. They wanted it all and they had very good reasons: the surrounding Arab nations supported the Palestinian aspirations and the Arab League decided to invade Palestine the moment the British left it in order to deny the Jews their promised land. U.N. Resolution number 181 was accepted on 29 November 1947, and the British were scheduled to withdraw on 15 May 1948, leaving the opposing sides half a year to prepare for the struggle. But it was not a peaceful period. Irregular Arab forces raided many isolated Jewish settlements, the roads were blocked, and it was dangerous to travel. The embargo on personnel reinforcements meant that many Jews that escaped the Holocaust in Europe were denied their future state, and the embargo on arms meant that the irregular, untrained, under-armed and highly motivated forces of the Haggana, as well as two other small movements (Atsel and Lechi), would have to stop the tanks of the Egyptian, Jordanian, Iraqi, Syrian and Lebanese armies with their bodies.

Almost a fortnight after Remez and Shamir handed over their work, on 2 November 1947 a meeting was called at the Jewish Agency building at Tel Aviv to decide upon the future of the plan. It was headed by Jacob Dori (later to become the IDF first Chief of Staff), Yiga'el Yadin (IDF second CoS) and Iseek Ashel. They decided to found Sherut Avir and adopted the Remez/Shamir work. Isheek Ashel was appointed as the SA commander, Alexander Zillioni head of staff, Ammi Copperman responsible for operations, Aharon Remez responsible for planning, Heyman Shamir responsible for training, Itschak Levy responsible for procurement, Yehuda Hollovich responsible for logistics, Shraga Yanai in charge of intelligence and security, and Yehuda Rabin chief engineer. Yehuda Arazi, a purchasing expert, was instructed to acquire aircraft and equipment. On 10 November 1947, Dori wrote Ashel on a formal paper:

Auster AOP.3 VQ-PAS seen in the Spring of 1948 at a forward airstrip after some heavy rains. (IAF)

"The Haggana's General Staff informs you of its decision to Found a regular air arm, commencing with this letter's date, which will be called "Sherut Avir." The plan (Remez and Shamir's) that you have introduced has been approved. You must forward at once your proposals as to appointments, budget and immediate realization of the plan."

The snow ball started rolling.

The Sherut Avir's staff offices were located at the Aero Club offices at 9, Montifury Street, Tel Aviv, and work was initiated immediately. All pilots available in the country were called to enlist, either by advertisements in the local press or by direct approach. Some were ex-WWII pilots with great experience; others were veteran PPL holders, and many had only a few dozen flying hours. But they all came, and it was soon realized that there were many more pilots than aircraft; a desired situation, but only up to a certain ratio. The pilots were categorized according to their experience; those who had a small number of flying hours were regarded as cadets, and it was decided to train them as a future reserve. The aircraft were also enlisted: two RWD 13s (VQ-PAL, -PAM); two Tiger Moths (VQ-PAT, -PAU); one Seabee (NC6731K); and two Taylorcrafts (VQ-PAI, -PAJ). The Zlin 12 (VQ-PAP) was not enlisted because it was an Atsel aircraft, and, although its pilot and registered owner, Benjamin Cahana, requested to join in with the aircraft, he was refused by the Haggana. The privately owned Bebe (VQ-PAQ) crashed on 29 November 1947 and its pilot and owner,

Scene at Lod Airport shortly before WWII.(Author's collection)

Ze'ev Levinson, was killed. The other aircraft previously mentioned were either withdrawn from use at an earlier date or were unserviceable at the time.

On 27 November 1947, the opening ceremony of the Sherut Avir (SA) took place at the Unition Trade House at Tel Aviv. Nine pilots, with some 120 flying hours to their credit, stood on one side, and about twenty cadets on the other. The SA missions were listed as liaison, transport, intelligence gathering, escorting vehicle convoys and training new pilots. It was all very promising, but there was one major shortcoming: the aircraft were based at Lod airport under British control and in the middle of an unfriendly Arab-populated area. All pilots had to fulfill flight plans that were approved by the British authorities; moreover, since many Arabs worked in the airport, and in fact even guarded it, it was feared that the aircraft could be sabotaged or set ablaze, and the SA's first operation was the transfer of the fleet to Sde-Dov airport and the creation of the Tel Aviv Squadron.

Sde-Dov had not been used since Palestine Airways ceased operations in 1940. During WWII the area became a complex of British army camps, and the single runway (10/28 heading) was covered with sand. Under the cover of Aviron and the Aeroclub, SA approached the British authorities and asked permission to reopen Sde-Dov for civilian use. Permission was granted, but the British demanded that each flight should land at Lod for the flight plan procedure and upon completion of the flight, land again at Lod!.

Nevertheless, it was a major breakthrough since for the first time the Jewish air arm could operate from a Jewish zone on the outskirts of Tel Aviv. As time passed the restrictions were ignored and the aircraft were based at Sde-Dov and operated at day and night without passing through for the flight plan formality. The transfer was quickly organized as a planned operation that was named Operation "Tsipora" (a biblical figure), and on the early morning of Saturday, 13 December 1947, Yehuda Hollovich led a group of cadets and mechanics to clear the sand from the old runway. They worked until noon and were helped by sympathetic British soldiers from a transport unit that used the runway as a parking ground. The trucks were moved, the sand was cleared, and a windsock was installed in anticipation of the first aircraft arrival. While Hollovitch's group was working, a group of pilots led by Emmanuel Tsur, Aviron's chief pilot, gathered at Aviron's offices at Echad Ha'am 70, Tel Aviv. At 1600 hrs in the afternoon, they went to Lod in an armed convoy, since the road passed through hostile Arab area, and climbed into the nine serviceable aircraft. The RWD 15 (VQ-PAE) was left behind in the Aviron's hangar because it was not serviceable, and, as originally feared, on the night of 13 April 1948, was set on fire by the Arabs. The two DH-82s, RWD 13s, Taylorcrafts, as well as the Auster, Seabe and Dragon Rapide, landed safely on the 300-meter runway, and the scene was set for the birth of the Tel Aviv Squadron.

In accordance with the Remez/Shamir proposal, the existing fleet of light-planes was organized as a squadron under the direct control of the air staff. The aircraft were parked beside the runway on PSP (Pierced Steel Planking) sheets that were laid over the sand, and a small tower painted in black and white checkerboards was raised nearby. The whole area was a closed British military zone, and the crews had to pass British checkpoints on their way in and out; and since one of SA's main missions was to supply small arms and ammunition to remote, isolated, and besieged points, the crews had to smuggle them into their aircraft through the British guards and under the constant tension of being caught. In order to improve the SA capabilities, many airstrips were constructed at isolated settlements, especially in the Negev, but also in northern Israel (the Gallelee). Naturally, those strips were hazardous because of short lengths and dangerous topography, and many were under constant sniper fire. In places where it was impossible to erect a strip, the supplies were either dropped or parachuted according to the surrounding area of the settlement, the type of supply, and the availability of parachutes. To many isolated points the small aircraft with its daily cargo of mail, newspapers, orders and urgent supplies was the only link to the outside world.

The first pilots of the Tel Aviv Squadron were Ezer Weizman and Moddi Allon (both ex-WWII RAF fighter jocks), Misha Kinner (ex-Soviet AF fighter pilot) and Nachom Efrat, Pinchas Ben Porat, Jacob Ben Chaim, Elli Eyal and David Shprinchak (PPL holders from the Palmach's Flying Platoon). It only shows the spirit of the period that they decided to choose the squadron commander through democratic elections each month!. The first to be elected was Elli Eyal—certainly not the most experienced pilot. Eyal was also the last to be elected since the staff disapproved of such advanced concepts, and his successor, Boris Senior, was appointed in the familiar military procedure by the staff. From this group of pioneers, three were killed in the coming months, but the rest carried on and many others joined in. It was well known that air power was a decisive force in modern warfare, but no one even considered light aircraft armed with hand-held machine guns and hand grenades as air power. No one except a young flier named Pinchas Ben Porat. He learned to fly with Aviron in 1940, and later with only 150 flying hours to his credit, in January 1944 he became the first commander of the Palmach's Flying Platoon. He was sent to England for a year for further training and gained there a commercial license, and when he returned in late 1947, he immediately joined the SA. During May 1948 he was among the small group selected for S-199 conversion, but broke his arm in a crash landing and was posted to fly transports and later B-17s. After the end of the Independence War, he was the first Israeli instructor in the Flying School, commanding the primary stage, and was later posted as the B-17 squadron commander as a RAV Seren (Major). In 1950 he left the Chel Ha'avir and joined El-Al (Israel national airline) as a Lockheed Constellation copilot. On 27 July 1955, 4X-AKC, an El-Al Constellation, was shot down by Bulgarian MiGs who mistook it for an American intelligence gathering aircraft. All aboard were killed, including the copilot, Pinchas Ben Porat.

On 17 December 1947 Ben Porat was sent to provide support to Nevatim, a remote settlement in the Negev under Arab irregulars attack. He landed there, took a "gunner" with a Bren, and performed "Close-Air Support." Nevatim was saved and Ben Porat pioneered the use of air power as a decisive factor in Haggana—and later IDF—tactical doctrines. Naturally, the British heard about the incident and warned the young fliers, but Ben Porat's example was followed and a small, noisy airplane dropping hand grenades and firing hand-held machine guns became a common scene during the following months. This action had such an impact that Ben Gurion himself invited the young aviator to report him and wrote in his diary on 18 December:

"...Pinia (nickname of Ben Porat) reported to me: yesterday at 0800 he flew from Sde-Dov to Lod with Dr. Polack and from there to Bet Ashel. Arrived at 0920. Discovered that the casualties were evacuated to Beylinson Hospital, near Petah Tikva. He was told that since the morning Nevatim was under attack and there were casualties. A British police officer arrived from Be'er Sheva and asked Ben Porat for his name and PPL. Pinia announced that he was sent by Aviron and the doctor by Magen David [the Israeli equivalent of the Red Cross]. The officer left for Nevatim but returned after half an hour (a distance of eight kilometers) and reported that Nevatim was surrounded by about 200 Arabs and the road was blocked.

Pinia took hand grenades, a machine gun and a lad, removed the door of his aircraft and flew to Nevatim at 1050 at an altitude of 700-1000 meters. He saw several groups of Arabs in canyons, dropped hand grenades and fired with the machine gun, he witnessed an Arab collapsing and landed on 1120 nearby the settlement, was airborne for 40 minutes. He left the lad and the machine gun at Nevatim. A formation of five/six British aircraft overflew the place at 1150 hrs.

He evacuated the two casualties to Tel Aviv in two flights.

When he was at Bet Ashel his aircraft was fired at from the Be'er Sheva area."

Scene at Lod Airport shortly before WWII.(Author's collection)

The growing workload imposed on the Tel Aviv Squadron was to such an extent that more aircraft were badly needed. By 24 December the SA monthly budget was 3.314 Palestinian Pounds, it had 66 men, and its aircraft logged some 500 hours per month. The Taylors, with only a 55HP engine, were of little operational use and were mainly flown by the "cadets" to gain flight hours and experience. The Rapide was mainly operated by Aviron for VIP flights, and it seems that the Seabee was not operated at all during that period. The RWD 13s, Austers and Tiger Moths performed all the missions, and reinforcements were badly needed. The outstanding effort to purchase arms all over the world will be treated in the next chapter, but the equipment that enabled the SA to dramatically expand its operations was purchased in Palestine. Since the end of WWII the RAF had operated a variety of aircraft on surveillance patrols from its bases in Palestine to prevent Jewish refugees from reaching their promised land, and among the units committed was No 651 Squadron, operating Auster AOPs from Lod. Towards the end of 1947 the squadron was disbanded and its aircraft auctioned. Under the cover of Aviron, SA negotiated their purchase, having been stored at Tel Nof air base. It was found that from the 20 Austers offered, only eight were in reasonable condition, while the others were regarded as junk. Nevertheless, Aviron won the auction on 14 January 1948 and Alexander Zillioni accompanied an armed convoy from Tel Aviv to Jerusalem to close the deal. The British demanded 10.300 PP while Aviron offered 7.500 PP. The agreement was on 8.500 PP for the 20 aircraft and a single unserviceable Airspeed Oxford. Once again Yehuda Hollovich was responsible for the logistic side of the operation, and in two days' work the Austers were transferred as truck loads from Tel Nof to the old wine cellar of Sharonna on the outskirts of Tel Aviv. The Oxford was left at Tel Nof. The Arab newspapers published the story and the British Government ordered the authorities in Palestine to cancel the deal immediately, but it was too late; the Austers were well hidden at Sharonna. The British authorities thought that not only were the aircraft useless, but even if brought to flying status the Austers still lacked civilian registration. They were wrong. Under the leadership of Yehuda Rabin a small group of dedicated mechanics performed the impossible task; at least 15 Austers were reconditioned. On early March 1948 the first refurbished Auster was transported to Sde-Dov, and the next difficulty, civilian registration, had to be overcome. The answer was simple: SA already operated a legal Auster J1 and two Taylors that resembled the Auster to the distant observer, and it was decided to paint the illegal Austers exactly in the colors of the legal Auster and Taylor. Nine AOPs were "registered" as VQ-PAI, two as VQ-PAJ (AOP5) and two as VQ-PAS (AOP3), and each aircraft was given a number (for example VQ-PAI/9) for identification and for maintenance schedules. The mechanics had to remember to park duplicated aircraft side by side, and many embarrassing incidents occurred when pilots landed at British-controlled airfields only to find their aircraft's "twin" parked on the ramp, or pilots were asked by the authorities to show their aircraft logbooks for inspection. The British authorities were either fooled by that childish trick or just ignored the whole affair, but SA could expand its operations, and, as more and more refurbished AOPs were taken on charge, two sectorial units, the Negev Squadron and the Gallilee Squadron, were formed.

On 4 January 1948 Ammi Copperman, head of operations in SA's staff, instructed Elli Eyal, Tel Aviv Squadron C.O., to qualify all pilots on all types and to split operational flights between all pilots on an equal basis. It seems that only a small number of experienced pilots undertook all operational flying, and the SA's staff objected to that trend, although it also seems that the staff objection had little effect on the Tel Aviv Squadron. In almost fifty years of perspective, it looks like the SA's staff was mainly concerned with what could be termed as "strategic thinking", i.e.: the relationship between the SA's staff and the Haggana's staff (or later Chel Ha'avir's staff and the IDF's staff), and was little involved with what was going on in the "field" where the eager young pilots of the Tel Aviv Sqn.. performed wonders considering their equipment. The importance of SA even in this early period of its life can be exemplified by the Kfar Oriyah incident on 11 January 1948. The small settlement commanded by Shlomo Lahat was attacked by large Arab Forces. Towards 1500 hrs, after hours of fierce fighting, they had almost run out of ammunition, when suddenly a small, noisy aeroplane came out of the blue and parachuted sub-machine gun and rifle bullets. Kfar Oriyah held on.

During all that time many pilots that were not founder members of SA joined the service, among them Daniel Buckstein, who gained his PPL with Aviron and in 1945 joined the RAF. His first operational flight was to drop supplies to Gosh Etszion on 19 January 1948 from an RWD 13. Gosh Etszion was a group of four besieged settlements in the heart of an Arab populated area, midway between Jerusalem and Hebron. The Gosh, as it was known, suffered a severe attack by irregular Arab forces from the neighboring villages and cities on 14 January, and it was feared that another attack would take place the following dawn. On the evening of 14 January the Haggana staff decided to parachute weapons, ammunition and medical supplies to the Gosh early the next morning since there was not any landing strip at that mountainous area. On the same evening "operational alert" was called at Tel Aviv Sqn.. and the pilots were briefed by SA staff officers. It was the first large-scale, planned operation by the SA by the standards of the day. The plan was that early next morning (15 January) Boris Senior and Elli Eyal would take off from Sde-Dov in a Tiger Moth armed with a Bren machine gun and hand grenades to cover the supply aircraft, while Ezer Weizman and Itschak Henneson would be in an Auster and Jacob Ben Chaim and Pessah Tal in an RWD 13. Each aircraft was crewed by two pilots: one flying the aircraft and the other dropping the supplies through the door. Since there were no parachutes the supplies were to be dropped from very low altitude, tied to car tires and at nearly stalling speed. Hand grenades, for example, were dropped without detonators to lower the risk of explosion and the detonators were dropped separately. If the dropping action would be discovered by the British authorities the aircraft should divert to Gazza East, a deserted landing strip near Be'erot Itschak (a distant kibbutz), in order to clear all evidence before proceeding to Sde-Dov.

Also at Be'erot Itschak, Eddie Cohen would wait in a Taylor with fuel jerrycans to perform refueling if necessary. There were no communication methods, and the plan had to be executed according to the previous planned timetable. The supplies were smuggled to Sde-Dov through the British checkpoints in taxi cabs during the night. Before dawn the Tiger took off. In the air Eyal tried the Bren—it fired a short burst and stopped. Only later it was revealed that the oil had frozen and the mechanism stuck. Anyway, the Tiger was not needed since no Arabs were present in the surrounding area and the dropping action was performed without any trouble except for a British Auster AOP that watched the action. Ezer Weizman claimed that some of the packages were completely destroyed, but the defenders greatly appreciated the effort and the supplies that were not damaged.

Since they were spotted by the British Auster they flew to Be'erot Itschak, cleared all evidence, refueled and continued to Sde-Dov. Upon arrival to this airfield they were informed that the British had approached the Haggana's staff and demanded that the pilot who took off under the cover of the Aero Club and dropped supplies to Gosh Etszion turn himself in since the British pilot only saw one aircraft. The dilemma now was whether to turn one of the pilots in or ignore the British demand which, in that case, might close the Aero Club and thus badly affect SA's operations. It was decided to turn in a pilot and the one chosen was none other than Ezer Weizman. His uncle was Chaim Weizman, one of the most important Jewish leaders of this century and later to become the first president of Israel. Chaim Weizman's son, Michael, had been an RAF pilot during WWII, and in February 1942 was listed as MIA over the Bay of Biscay while flying his Whitley bomber. Ezer Weizman gained his PPL in the summer of that year, joined the RAF as a truck driver in September 1942, undertook a flying course at Rhodesia in early 1945, and served as a Thunderbolt pilot in India. In June 1946 he went to study at the Aeronautical College at London, but returned to Palestine in the summer of 1947 and joined the SA. Weizman was chosen because of his "perfect" past record, while his cover story was that he heard about the attack on Gosh Etszion, bought medical supplies, leased an aircraft from the Aero Club and dropped the supplies to the Gosh. Weizman was arrested for two days and released. It is almost ironic that exactly a year later, on 7 January 1949, Weizman, this time flying a Spitfire, claimed an RAF Tempest in one of two air combats which took place that day between the Chel Ha'avir and the RAF in northern Sinai. Later on Weizman was Chel Ha'avir commander during 1958-1966, Minister during 1977-1981, a member of the Parliament (Knesset), and actually President of Israel.

SA operations were saved and it was on that same month, January 1948, that Yehuda Arazi, the Haggana's procurement expert sent previously to the U.S. to buy arms, returned for a working visit. Arazi already had three Lockheed C-69 Constellations and ten Curtiss C-46 Commandos under refurbishment in the U.S., and was called to discuss the possibility of using them to transport arms from Czechoslovakia, one of the few countries that was willing to help the future state of Israel. The plan was to use a legal civilian

Palestine Airways Ltd. Shorts S.16 Scion pictured in March 1938 at Tel-Aviv airport, later named Sde-Dov. (Author collection)

airline to be based somewhere in Europe, preferably in Italy, thus establishing an air link with Israel to deliver the arms. Seven locations were chosen as secret airfields, two at Hertzlia and the rest at Kfar Shmaryaho, Kyriat Shaul, Aben Yehuda, Ma'abarot and Sde-Dov. All locations were within Jewish populated areas, and in the first six runways were straightened by tractors, while at Sde-Dov it was lengthened; dimensions were between 700 and 1000 meters long and 50 meters wide. However, none of these landing grounds was ever used by Arazi's aircraft. By the time they arrived in Europe, excellent vacated ex-RAF airfields were already in regular use by the SA. Some of these six airstrips (excluding Sde-Dov) were later used regularly by the Chel Ha'avir; Hertzlia was the only fighter base during the summer of 1948, Ma'abarot was the Central Maintenance Unit base until the autumn of 1948, and Aben Yehuda was briefly used by a single B-17 in July 1948.

Although Sherut Avir operations spread all over Palestine, its activity during February 1948 focused on Gosh Etszion. The besieged settlements were completely isolated except for the air link. However, it was found out that in the dropping method, a large percentage of supplies were severely damaged, while the parachuting method lacked the required accuracy and many packages fell beyond the reach of the defenders. It was decided to build a landing strip at the Gosh despite of the hazards—mountainous ground some 1000 meters high with surrounding hills and constant small arms fire from the besieging Arabs. On Saturday afternoon, 7 February, Itschak Henneson pioneered the 200 meter strip in an Auster with a sack of mail and newspapers. Elly Ayal, for example, made three round trips to the Gosh on 21 February carrying ammunition. A totally different method to supply the Gosh was also tried during January, but ended in a disaster. A unit of 35 men tried to reach the Gosh through the Arab's lines, but was discovered and completely wiped out after a bitter fight that lasted all day. It was decided to retaliate and attack the village of Sassa. The operation was carefully planned, the intended commander of the operation was flown over the target in one of SA's aircraft and the pictures which were taken on the same flight by a hand-held Leika camera were used by the planners and for the briefing. The force blew up 35 houses in the village.

The increasing demand for aerial support necessitated the formation of sectorial flights. The limited speed and endurance of the light-planes meant that they had very short time on station when flying from Sde-Dov to the Negev, for example. It also meant that the response time to urgent requests was unavoidably long. The infusion of Auster AOPs to the SA cleared the way to the activation of more units, and on 12 March 1948, the Negev Squadron came into being. Ezer Weizman was appointed its commander, and the base was established at Nir Am near the Negev Brigada's HQ. The other pilots were Elly Eyal, Nahchom Biran and Jacob Avisar, while three mechanics were detached from the Tel Aviv Sqn.. to give line maintenance. The Negev Sqn.. was directly operated by the Negev Brigade and relied on the Tel Aviv Sqn. for routine maintenance; the Auster in need of maintenance was flown to Sde-Dov and, if in need of major or lengthy repair, was usually replaced by another aircraft to maintain the squadron status of two operational aeroplanes. The Negev Sqn. missions were daily routine observation flights to check the water pipe which supplied the dry Negev its water and was frequently sabotaged by the Arabs; daily mail and newspaper flights were made to all the points in the Negev. Some of the settlements had landing strips, while on the others the supplies were either dropped or parachuted. The Negev Sqn. also supplied the settlements with medical supplies and badly needed items, such as flour. Other missions were the direct support of the Negev Brigade units and medical evacuation of wounded to hospitals in Tel Aviv. Relieved from responsibility to the southern part of Palestine, the Tel Aviv Sqn. concentrated on the central part. A typical flying day was performed by Daniel Buckstein on 22 March 1948. On his first flight that day he dropped medical supplies to kibbutz Nitzanim, south of Tel Aviv, and on the second mission he flew to kibbutz Beit Ha'arava—on the shore of the Dead Sea—to pick up one of its defenders for an observation of the area. Gosh Etszion was still attracting a lot of missions from the Tel Aviv Sqn., and on 21 March an operation to fly a badly needed enforcement platoon into the Gosh was initiated. This may seem an easy task today, but it was a major operation for the single-engined and worn out Sherut Avir aircraft that could only carry one or two passengers on each flight. On the first day, nine men were transported in seven sorties. Each round trip took an hour and half, and the aircraft also transported the weapons and ammunition of the platoon. The "airbridge" set up by the Tel Aviv Sqn. could not deliver an adequate amount of supplies needed to ensure the survival of the Gosh, and the Haggana decided to organize an armed convoy that would force its way from Jerusalem into the Gosh. It consisted of 37 trucks and 14 makeshift armored cars with armed escort and an observation aircraft to patrol the road in front of the convoy.

The convoy left Jerusalem at 0830 hrs in the morning of 27 March, and three hours earlier Daniel Buckstein, along with a radio operator, took off from Sde-Dov in an Auster and surveyed the route. He reported to the convoy that the road was free of barricades and ambushes and returned to the base. He later took off in a Taylor and saw the convoy crossing the hostile city of Beit Lehchem and entering the Gosh at 1000 hrs. Buckstein landed there and one hour later took off to survey the return route. He reported some three barricades near Shlomo's pools, where he was welcomed by rifle fire, and another two barricades still further north and many armed Arabs reaching the Jerusalem-Hebron road at the Beit Gal'lah area. The convoy lost its surprise factor and it was obvious that it would have to fight its way back to Jerusalem. Buckstein returned to Sde-Dov and passed on the convoy's commander demand for aerial support. The convoy managed to break through the first few barricades, but was finally stopped near Nebbi Daniel, and in that place Itschak Henneson observed the besieged convoy at 1400 hrs. Henneson confirmed radio contact with the convoy and received targets to attack. He dropped two bombs and fired his Bren Machine gun until it stuck, when he then landed at nearby Gosh Etszion, supplanted the jammed Bren for a working one, and returned to support the convoy. Before leaving the place for Sde-Dov, Henneson noticed an RAF Auster flying in the area spotting the battle. Despite the risk of British retaliation, the support of the besieged convoy was regarded as top emergency and all other considerations were left aside. During the afternoon Giddeon Elrom dropped one home-made 25-kg bomb (these bombs were steel tubes filled with HE and before dropping out of the aircraft's door the "bombardier" removed the safety pin of the fuse in the same manner as operating a hand grenade. The bomb was designed to explode on the ground if dropped from 2000 ft and was introduced into service during March, replacing the hand grenades) and fired fifteen Bren's magazines. Sherut Avir operations continued into the darkness for the first time. As early as 14 March Alexander Zilloni, SA's head of staff, sent orders to initiate a night flying training scheme for all the pilots, but it seems that the order was not pursued and many SA pilots lacked night flying capability. Pessah Tal, an ex-WWII USAAF B-25 pilot in the Pacific theater was the first to fly on the full moon night of 27/28 March and, despite of the heavy anti-aircraft fire, dropped ammunition to the convoy, but on landing at Sde-Dov he overran the runway and wound up in the sand without damage to the aircraft. Next to operate that night were Emmanuel Tsur and Daniel Buckstein flying the Dragon Rapide. They reached the scene just before dawn, but still under the cover of darkness. They too faced heavy fire and dropped three food packages marked with red flashlights to enable their collection by the convoy's defenders. The packages were lost, but the Rapide's pilots refused to go lower for more accurate droppings because of the AA fire. The convoy situation was so desperate that its commander insisted on further dropping attempts. The next three packages landed within 100 meters of the convoy but could not be retrieved, and from the final batch of three packages, only one reached the convoy and was collected by its defenders. The plan for the next morning (28 March) was to attempt once more to drop supplies to the convoy by the Dragon Rapide, only this time enjoying daytime accuracy and visibility and having an armed escort aircraft to act as a decoy. Yesha'ayaho Bodilovski and his bombardier, Matit'yaho Socenic, took off from Sde-Dov at 0650 hrs with eight bombs, a Bren and ten magazines, but the Dragon Rapide did not show and thus they attacked the besieging Arabs and they too noticed an RAF Auster spotting the area. At 0850 hrs Peretz Grosser and his bombardier arrived at the scene, but were instructed not to attack since a negotiation was in course with British forces regarding the evacuation of the convoy and they landed at the Gosh and waited there for further orders. The negotiation failed, and at 1510 hrs Shamuel Widliss and Tsvi Zibbel took off from Sde-Dov to drop food to the convoy.

In their second pass a bullet struck the airspeed indicator and Widliss was wounded by glass fragments flying at his chest and face, so they immediately returned to Sde-Dov. Meanwhile, at 1520 hrs, Grosser took off from the Gosh and attacked with his eight bombs, of which four failed to explode, thus showing the poor reliability of these homemade weapons. Grosser also faced heavy small arms fire, and the engine's support was damaged and he was forced to return home. In the following night Pessah Tal tried once more to drop supplies, but due to heavy Arab fire and engine troubles he was forced to abandon. Next day the convoy was evacuated under British cover (the British took all the convoy's weapons, but did not do the same with the Arabs), and SA'soperations in support of the so-called Nebbi Daniel Convoy could be summed up as follows: 14 pilots and six aircraft took part and executed 37 sorties (35 flying hours) including ten night sorties (13 flying hours). The total number of bombs dropped was 33 in ten bombing sorties, and although the actual damage suffered by the attackers from these bombs can be considered to be very small due to lack of accuracy (no bombing sights were used, and the bombs were thrown out of the window, or through the open space left by deleting the side door, by the bombardier according to the pilot's orders which were merely based on instincts and experience) and reliability (some of the bombs exploded in the air without causing any damage, while others did not explode at all) the real achievement of the bombing sorties was the serious deterioration of the morale among the Arabs on one hand and the rising of the convoy defenders' morale on the other hand.

Towards the end of March 1948, enough operational experience had been gained by the Sherut Avir and its professional staff members demanded changes. Up to that period, and in fact in many years to come, the SA was regarded by the Haggana's staff as a technical service which supplied the means to enable the latter's staff operations department to plan and execute operations. The SA's staff did not participate in the planning stage. A liaison officer from the operations department passed to the SA's staff the operational requirements. For that reason it was considered by the Haggana's staff that the SA's commander must not necessarily be an aviation expert and Iseek Ashel, although a veteran Haggana member, had no aerial background. On the other hand, the young SA's staff officers had aerial background, especially in the RAF and USAAF during WWII, and demanded that the SA should participate in the planning stage since only they could advise the planners on how to use best the existing airpower. Moreover, the rapid expansion of the SA induced logistic difficulties, and hence, in a meeting at Yehuda Hollovitch's house at Tel Aviv, the SA's staff group decided to approach David Ben Gurion and demand to set up an HQ that would directly command the SA and would work in cooperation with the ground forces staff, and to appoint a dominant figure that would resume responsibility on the logistic side of SA's operations. Only in early March 1948 did Ben Gurion partly accept the staff demands, but already in early April Iseek Ashel left the Sherut Avir and for the following month it was run by the staff and the sectorial squadrons without an official commander.

On the night of 31 March 1948 an event of major importance took place at the deserted RAF airfield of Beit Daras when a Douglas C-54 Skymaster, wet-leased from U.S. Overseas Airlines at Geneva, landed there with the first cargo of arms from Czechoslovakia. The four-engined transport was navigated by Mordechai "Mocca" Limon, a veteran seaman later to become the Israeli Navy commander, who was accompanied by Ammi Copperman, SA's operations officer. The navigator of the original American crew was an American Jew, Seymore Lerner, who later volunteered to Chel Ha'avir and was killed while ferrying a Mosquito from Canada in the early fifties. A special HQ was set up for Operation Hassida (Stork) as it was known: Aharon Remez was the commander, Yehuda Hollovitch responsible for transport, logistics and electronics; Yehuda Rabin for mechanical servicing and refueling the C-54; and Peretz Rosemberg for communications. The Beit Daras airstrip was originally constructed by the RAF during WWII as an auxiliary site to the nearby Hatzor air base and was at the time deserted except for a small force of Arab policemen which were expelled from the strip by Haggana forces on 29 March. The Givatti Brigade was responsible for the protection of the area and the unloading of the

Members of Aviron's flight course pictured on 10 June 1943. In the foreground is a RWD.13 (SP-BFM) and a RWD.8 in the background. (Author's collection)

arms. On the night of 29/30 March the electrical strip lighting, a communications truck and refueling trucks, which in fact were regular-decked trucks with fuel barrels and hand pumps, were all brought to the site. The communications procedure that was supposed to begin at 2000 hrs on 31 March commenced with a slight delay, but the operation proceeded according to the plan and 200 Czech rifles, 40 German MG 34 machine guns and 150.000 7.92 mm rounds that were badly needed by the Haggana forces were quickly unloaded to enable the C-54 to take-off before dawn. As daylight broke all the forces and evidence were gone, but as the Skymaster landed at Praga, its crew was interrogated by the local U.S. Embassy officials and it was understood that future operations would have to rely upon Arazi's aircraft of which the first, a Curtiss C-46 Commando registered NX67923 and operated in LAPSA colors (see next chapter), was already in Italy from 9 March 1948, but was not used to carry arms from Czechoslovakia at that stage in order to not risk the export from the US. However, on late March 1948, that Commando was used for some peculiar operations. The Haggana representatives in Italy were informed that a small Italian vessel named Lino was about to set sail from a Yugoslav harbor carrying 6.000 rifles, eight million 7.92 mm rounds, and a large number of hand grenades, as well as a load of explosives. The cargo was intended for Syria, and it was decided to do everything possible to deprive Syria from that cargo. But first Lino had to be located, and for that purpose the Commando and its crew of American volunteers searched the Adriatic Sea for hours from its base at Perugia. Once the ship was found, it was planned that the Commando would attack it with a home-made bomb thrown from its rear door and by firing at the ship with a 20 mm cannon through the same door. Fortunately enough for the crew it was found out by other means that Lino was still in harbor on 30 April and the plan for the aerial attack was abandoned. Later on the vessel was sunk by frogmen at the harbor of Bari in early April.

During the first half of April 1948, Jerusalem suffered from a critical shortage of supplies. The Haggana planned its first massive operation in order to open a corridor from Tel Aviv to Jerusalem through which the convoys, consisting of hundreds of trucks, would force their way. The operation was called Nachshon (Pioneer), and its realization could be partly attributed to Operation Hassida arms reinforcement. The forces were taken from all over Palestine, and thus the peripheral sectors were severely weakened, but the Haggana leaders accepted the risk. The Sherut Avir detached an aircraft to Nachson HQ at kibbutz Na'an that was mainly used for observation flights and intelligence gathering missions—usually an intelligence officer accompanied the pilot on these sorties. The Tel Aviv Sqn. also supported the operation from Sde-Dov mainly through bombing and dropping sorties, and for the first time formation bombing was attempted. On 2 April this unit performed five dropping sorties to supply the Ma'allah Ha'chammisa kibbutz roughly halfway between Tel Aviv and Jerusalem. The next day another seven similar missions were flown, and the Na'an strip was inaugurated by Modi Allon, who was the first pilot detached to the Nachson HQ in a Tiger Moth. Allon commenced the observation flights, but the next day, 4 April, all flights were canceled due to bad weather. On 6 April the first convoy to pass through the corridor that was opened after fierce fighting reached Jerusalem, and it was on the same day that the SA supported the forces in the Castel/Tsoba area. At 0500 hrs Pinchas Ben Porat bombed strongpoints in that area with three

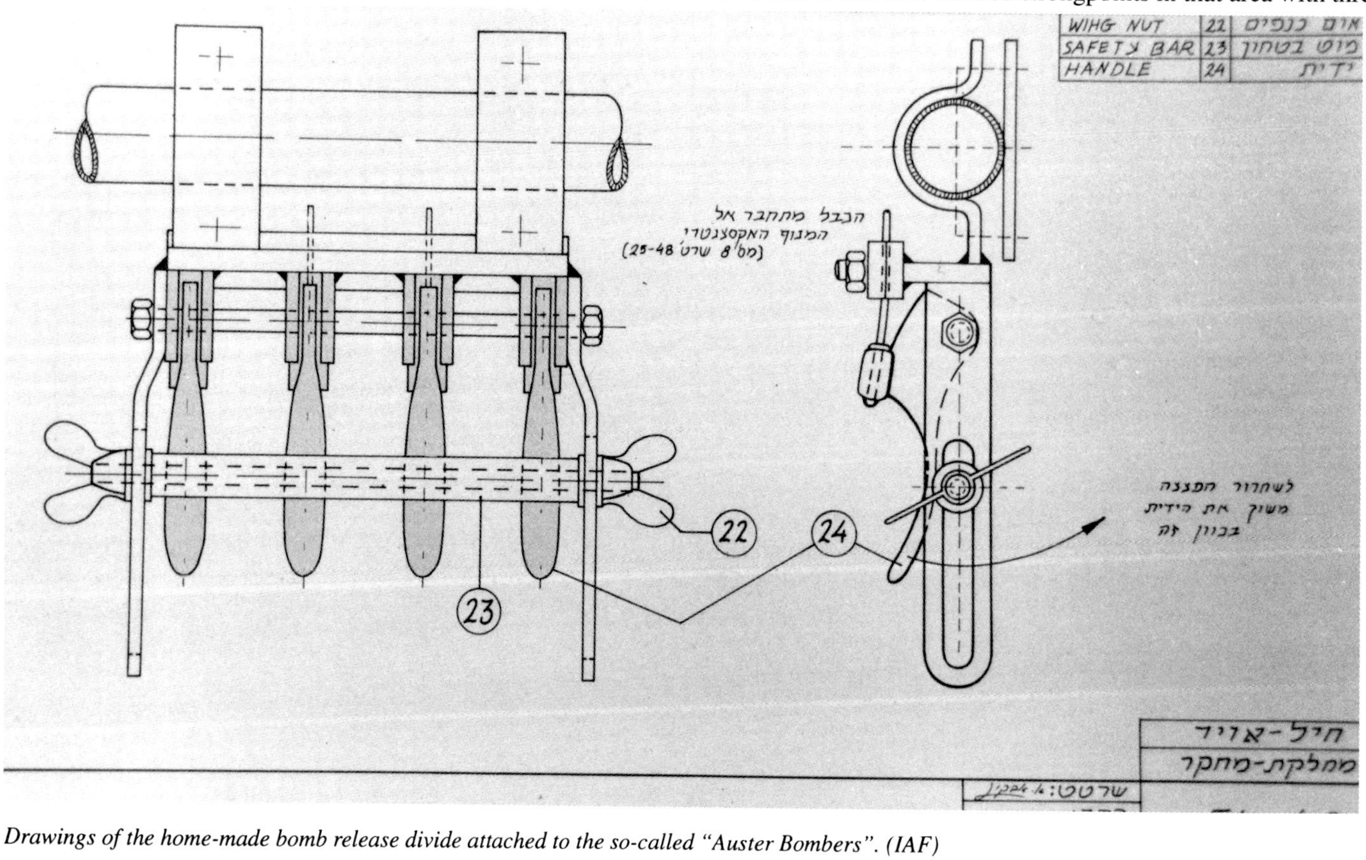

Drawings of the home-made bomb release divide attached to the so-called "Auster Bombers". (IAF)

20 kg and four 8 kg bombs; Jacob Ben Chaim replaced Ben Portat in that mission and also observed three barricades on the main road to Jerusalem near Motsa. The fierce fighting around the Castel key area continued, and on the night of 6/7 April at 0300 hrs Itschak Hennenson bombed Arab forces. The method employed was very simple; Hennenson dropped one 8 kg bomb on what he considered to be the targets, while the ground forces nearby saw the explosion and instructed him to drop the rest, another two 25-kg and three 8-Kg bombs, exactly at the same point. Hennenson returned to Sde-Dov and after a surprisingly long turn around of one hour he took off again and bombed the same target once more while employing the same method (the instructions were given through short range equipment). Roughly the same missions were flown during the next few days, and the pilots suffered from heavy small arms fire which restricted the minimum bombing height and affected accuracy. A typical Nachshon operation flight is narrated by David Spritsak, who flew an Auster on 9 April to escort a convoy:

". . . Taking off, turning to Holda (a kibbutz on the way to Jerusalem) I met the convoy on the main road near the entrance to Holda. There were seven buses and many trucks. The radio communication with the convoy was fine. The convoy waited for malfunctioning trucks. The head of the convoy reached Holda. At the same time a military looking car that followed the convoy in 300 meters distance turned back to Dir Moasieen where it met another military car that came from Jerusalem. It drove for Waddi Sarrar, waited for all the convoy to pass towards the Holda road and carried on until about one kilometer from the blown bridge of Waddi Sarrar. Three men jumped off, one in Arab looking clothes. They crossed the field towards Wadi Sarrar. We informed the convoy. At 1540 an armored car, built and painted like ours, was observed on the other side of the bridge. We informed the convoy and our mortars opened fire in its direction. It was not damaged, but immediately escaped. It may be that the armored car came to collect the three men previously. The convoy was all concentrated at Holda."

On the same day Moddi Allon was replaced as the pilot detached to the Givatti Brigade—which was responsible for the operation—by Giddeon Elrom, and on 12 April Tsvi Zibbel was also detached to help Elrom. The detached Tiger Moth was serviced at Na'an by two mechanics that were also detached from the Tel Aviv Sqn. for that purpose. Elrom was the pilot that located the battery of Egyptian artillery between Ma'allaeh Ha'chamisa and Jerusalem, and the target was given such priority that for the first time formation bombing was attempted. Three Austers were dispatched from Sde-Dov with Itschak Henneson leading, but the operation did not follow the planned course of events as can be seen by the reports of the three pilots:

ITSCHAK HENNENSON: ". . . Took off as the leader of a three aircraft formation, each one carrying three 25-kg bombs. One aircraft returned 30 minutes after take-off due to engine troubles. The radio communication among the aircraft was not good, and the operation suffered because of that reason. Two aircraft reached the target and dropped their bombs. I dropped one bomb on an Arab vehicle concentration—some eight cars—at the Arab village of Biddo. I dropped two more bombs at the eastern side of Beit Tsurif, where I noticed some vehicles, and possibly some artillery concentrations. At both villages I observed groups of men. The bombs were dropped from 2000 feet above ground level and all exploded. Heavy fire was encountered at both places."

SHAMUEL MIDLISS: ". . . I should have flown in a formation that Itschak Hennenson led. The powerplant of my aircraft lacked the power to maintain formation flight. I could not see Hennenson so I followed Pinni (Ben Porat) until I noticed that he was altering his course and I continued towards the target. After a few minutes Pinni said on the radio that he was returning to base because of mechanical failure. I tried to inform Itschak about my position all the time. I flew to Bir Zeid and patrolled the roads to Biddo. The communication with Itschak almost ceased, the radio sets were totally useless for aerial communications. I observed two concentrations of forces, one at Biddo and the other at Bir Ig'za. Because of the bad radio contact there were difficulties in the target recognition. Finally I dropped one bomb on vehicle concentrations at Biddo. It exploded. At that time the engine malfunctioned and I returned to base with the remaining two bombs."

PINCHAS BEN PORAT: ". . . I took off with three 25-kg bombs, a bombardier and a radio set. After 20 minutes I decided to return to base due to engine troubles."

On 14 April an interesting incident occurred to Itshak Henneson while flying an RWD 13, and this is what he says in his report:

". . . I delivered ammunition to Gosh Etszion. The engine refused to start and engineer Jacob Ment changed a magneto. At 1129 hrs I took off with a passenger. Immediately following take-off I had to make an emergency landing. The wheels, exhaust and engine casing were all damaged. With the help of people from the Gosh and engineer Ment, I brought the aircraft back to flying conditions next day at 1500 hrs. The emergency landing was due to loss of power immediately after take-off. I took off at 1615 hrs after the repair that included the gluing of a broken piece from the propeller detached and enormous vibration followed. I was afraid that the engine would break so I shut it down and made an emergency landing. With the aid of Jacob Ment I cut off the propeller to preserve symmetry and took off again, after ten minutes of circling above the Gosh I turned to Tel Aviv. Above the mountains the engine malfunctioned and I returned to the Gosh, leaving the repairs to the next day. After we changed a complete carburetor from another aircraft, the engine worked properly. At 1035 hrs I took off with the shortened propeller. After I circled the Gosh for ten minutes I continued to Tel Aviv and arrived at Sde-Dov at 1115 hrs."

On the night of 19/20 April the Palmach's third regiment attacked Nebi Yoshah, an ex-British police station on the Lebanese border, now held by Arab forces. The attack failed, and 22 of the attackers were killed. Pessah Tal was sent to bomb Nebi Yoshah on the morning of 20 April:

". . . After about 20 guys were killed I was sent to bomb Nebi Yoshah. It was a Tegart Fortress with a concrete ceiling, one and a half meters thick. I flew with a bombardier in an Auster and I encountered machine gun fire. I had four bombs and I made four passes. In each pass I dropped a bomb for better accuracy. In the fourth pass I received several hits in the windshield and the glass fragments struck my face. I was covered with blood. The oil tank was also damaged and the engine stopped. I had to perform an emergency landing in a field near kibbutz Holata and the aircraft was later repaired."

The next day, 21 April, Daniel Buckstein replaced Pessah Tal as Gallilee Sqn. commander since Tal, a very experienced multi-engined aircraft pilot, was required for a new mission: the establishment of a regular link between the Jewish zone of Palestine and Europe. All airlines had ceased operations from Lod, which was still Arab held, because of the dangerous situation, as airplanes had been fired at during final approach to Lod in the previous months. It was decided to lease a Douglas C-47 and open services to Europe from Haifa airfield, which was under British control until late May 1948, and also from Sde-Dov. The C-47 (ZS-AVK) was chartered from a South African airline, Pan African Air Charter (PAAC), by Boris Senior, who was at the time in South Africa for procurement purposes (see next chapter). Its pilot was a South African named Claude Duval, and Pessah Tal was appointed as his co-pilot. The C-47 commenced operations in late April especially for VIP flights to Europe, usually to Rome or Geneva through Athens, but also to supply besieged settlements within Palestine and especially Jerusalem.

As the road to Jerusalem was blocked since Operation Nachshon had ended, a small landing strip was erected near the present location of the Knesset. Jerusalem is located on a mountainous ridge, about 800 meters above sea level, and the landing strip was situated in a small valley surrounded by hills. The hazards also included unpredictable wind gusts and constant Arab sniper fire from the nearby neighborhood of Katamon. Pinchas Ben Porat inaugurated the new landing strip on 16 April with a Tiger Moth, and it was mainly used for VIP movements and to bring urgent supplies to Jerusalem. Because it was restricted to single-engined light-planes the supplies flown were mainly mail and ammunition, but the Jewish population in besieged Jerusalem was over 100.000 and food supplies were very low. Thus, it was decided to use the C-47 to drop supplies to Jerusalem, but the leasing contract included a clause that the crew would not fly to dangerous areas, although one can ague that all the country was one huge dangerous area at that time. Other difficulties were that British forces were still present at Jerusalem and that the C-47 was based at Haifa, also under British control. Thus, the supply flights to Jerusalem were flown at night after a fictitious flight plan to Cyprus was filled to satisfy the British authorities. Instead of flying to Cyprus the Dakota headed to Jerusalem and parachuted bread, water, weapons and ammunition to its defenders, returning to Haifa before dawn from the west to "follow" the flight plan.

On 25 April the battle for Jaffe, the Arab populated city neighboring Tel Aviv, was opened. After Arab snipers made life miserable at the Tel Aviv suburbs, the Atsel organization decided to outflank Jaffe and launched an attack on one of Jaffe's suburbs. The Atsel was assisted by the old Zlin 12 (VQ-PAP) that was flown from a strip at Tel Livinski, east of Tel Aviv, by Jacob "Jack" Ratoshniak, an ex-RAF pilot. The Zlin was based there since mid-April and was used during the combat for observation. As it was the first attack by Jewish forces on a large Arab populated city, the British responded with force. British artillery shelled Tel Aviv and Spitfire FR.18s attacked Jewish forces. The Spitfires were operated by the RAF from Ramat David by Nos 32 and 208 Squadrons, being mainly used to cover the British withdrawal through Haifa harbor until 23 May 1948, and both squadrons were destined to combat REAF and Chel Ha'avir Spitfires in the coming months.

As the Atsel was hard pressed, the Haggana forces joined the battle on 27 April and until 30 April Jaffe was entirely outflanked. Most of its 70.000 residents fled, and it was finally surrended on 13 May. As the Haggana forces joined the combat, so did the Sherut Avir, and mostly observation sorties were flown. On the other hand, for the Atsel-owned Zlin this was its last operation, as it was written-off in a heavy landing the following month.

In Early May 1948 Israel Amir was appointed as the Sherut Avir commander. Since early April, when Iseek Ashel left the service, the SA was run by the staff members, but it was quite clear that these men, who served as junior officers with the Allied air arms during WWII and had not even commanded small units such as squadrons, were not yet fit enough, or mature enough, to command the SA. The Haggana's General Staff realized that the SA staff could control operations up to a certain limit, but not logistics, administration and engineering, and for that purpose a veteran commander with good organizational capability was required. Amir's appointment was met with disappointment within the SA staff, who demanded a "professional", which meant a commander with aeronautical background and understanding. Amir, although a veteran Haggana member, had no aerial background at all. Chai Issaschar was appointed responsible to all the logistical needs of SA, and he was to work with Amir and relieve him of that burden. Issaschar's—who was under the direct control of the Minister (David Ben Gurion)—appointment was welcomed by the staff members, contrary to Amir's. To satisfy all, usually an impossible desire, a strange proposition was forwarded of which Ben Gurion wrote in his diary on 11 May:

". . . Israel (Amir) is appointed as the director of Chel Ha'avir (Ben Gurion wrote Chel Ha`avir and not Sherut Avir). Aharon (Remez)—or somebody else—is appointed as commander of the Chel Ha'avir. The commander is responsible in front of the director and the important appointments are proposed by the commander to the director for his approval. The director is responsible for all the needs of Chel Ha'avir so it can fulfill its mission. The command is by the commander. The director informs the commander about malfunctioning. If he proposes to relieve someone and the commander objects, then he brings it to the staff."

Such a command structure could hardly be expected to work, and indeed Amir insisted that if he had to serve with the air arm it would only be as the commander and without the "director" functions. Finally, Amir's ultimatum was accepted on 16 May and following his appointment as the commander of the Chel Ha'avir the staff changed a little. The original proposition in early May was: Alexander Zilloni Head of Staff; Joshua Gilloch Administration Officer and Aharon Remez, Planning and Operations Office. That proposition was short-lived, and on 16 May Aharon Remez was appointed Head of Staff. However, the main cause for which the staff members approached Ben Gurion about a month earlier, the relationship between the Haggana's General Staff and the Sherut Avir's air staff, was not dealt with and the situation was left without a change, which meant that the SA was still an executing force rather than a planning and executing force as its staff members desired.

The last fortnight before the creation of Israel on 15 May 1948 saw fierce battles as both sides tried to improve their positions towards the final British withdrawal and the Arab nations invasion. On the early morning hours of 1 May, the Lebanese army, supported by irregular local Arab forces, attacked kibbutz Ramot Naftali. Despite heavy AA fire the Gallilee Sqn. had attacked the Lebanese forces since noon. Moshe Peled performed three bombing sorties and Daniel Buckstein another one. The next day the Lebanese forces broke into Ramot Naftali and it was not clear if the place fell or if it was still fighting. Peretz Grosser was sent to clarify about that point—he flew so low that the Lebanese were firing at him from above, and he dropped a package which contained colored flags, along with instructions and a note that informed the defenders, if any were left, that a reinforcement would reach them that night. The instructions were to wave the red flag if any defenders were still fighting, the white flag if they faced ammunition shortages, the yellow flag if they needed medical supplies, and so on. Needless to say, with such imaginative minds around, Ramot Naftali survived.

A week later, on 8 May, Buckstein performed eight sorties with senior ground forces commanders, all in the Fairchild, in which Arab concentrations in the Gallilee, southern Lebanon and the Golan Heights in Syria were observed in order to gather information towards the coming invasion. On the same day Operation Maccabbi (after the name of a killed commander) commenced at the central sector. Its aim was roughly similar to Operation Nachshon, but it was far less successful. The missions tasked to the Tel Aviv Sqn. were to support the hard-pressed forces, but bad weather and cloudy skies with restricted visibility affected operations. On 9 May a Tel Aviv Sqn. aircraft was to bomb Beit Machsir, an Arab populated village under attack by Haggana forces, but it could not find its assigned target due to clouds. The next morning two aircraft were sent to attack the same targets and Boris Senior, the Sde-Dov commander, was the first to take-off in an Auster. Just a few days earlier Senior returned from a procurement mission in his native country, South Africa, in a shining new Bonanza that he flew through Africa. Senior's aircraft was damaged by AA fire and he was forced to execute an emergency landing at kibbutz Kiryat Annavim. The second aircraft was a U.S. registered, Canadian manufactured, Noorduyn Norseman. It was one of 20 bought by another procurement agent, Freddie Thorsh Fredkins, from surplus USAAF stocks in Germany, and the first two arrived at Sde-Dov in early May. The Norseman was a real giant compared to SA's aircraft of those days and its carrying capacity was equal to more than five Auster payloads. The Norseman was flown by Yariv Shainbuim and its crew consisted of two bombardiers and two ground forces men that were to guide the aircraft to its target despite the bad visibility caused by low clouds. The first home-made 100-kg bomb was also carried by the Norseman, which was now ready for its take-off run. On the same morning Daniel Buckstein arrived from the Gallilee in an Auster that required maintenance. Buckstein asked, and was granted permission, to join the Norseman as a copilot. The Norseman reached the target's general area, but crashed there due to an unknown cause. A ground unit tried to break through to the wreckage, but suffered casualties and had to retreat. It was the SA'a first fatal loss, and later the wreckage was found and the P&W R-1340 engine can still be seen on Chel Ha'avir memorial at Ha'tayasin ("Pilots") Mountain west of Jerusalem.

On 12 May the Jordanian army, aided by local Arab forces, attacked Gosh Etszion. By the afternoon of that day the defenders'situation was desperate and they asked for urgent air support. Aircraft from the Tel Aviv Sqn. appeared and dropped ammunition to the defenders and bombs to the attackers but their effectiveness was poor. On the afternoon of the following day an aircraft from the same unit observed that Jordanian armored cars had entered Kfar Etszion. Its defenders—127, including 21 women—were massacred by the local Arabs, and the next day the three remaining settlements in the Gosh, Masho'ot Itschak, Revadim and Ein Tsurim, surrended, and their defenders were taken POW by the Jordanian army. The demise of Gosh Etszion was a major blow, and it was decided to evacuate a few besieged settlements that were in real danger of falling into Arab hands. Kibbutz Beit Ha'arava and the Death Sea Works at Calia, just south of Jericho, were among those settlements, and the Dakota was sent on 14 May to evacuate women and children from Calia. The Dakota had engine trouble and returned, so a Dragon Rapide was sent instead, but naturally it was unable to evacuate all the women and children of Beit Ha'arava.

Following the Norseman's crash, on 10 May Moshe Peled was appointed Buckstein successor as the Gallilee Sqn. commander only two days after Elli Eyal succeeded Ezer Weizman as the Negev Sqn. commander. Weizman, a temperamental figure, was one of a group chosen for conversion to the Avia S-199 fighters acquired in Czechoslovakia as the first warplane of the yet-unborn Chel Ha'avir. The other members of that group were Moddi Allon, Eddie Cohen, Pinchas Ben Porat, Itschak Hennenson., Nachman Mei'ri, Misha Kenner and Jacob Ben Chaim. Aluf them were veteran SA members, and they were joined by two Jewish volunteers from the U.S.: Lou Lennart, an ex-USMC fighter pilot, and Milton Rubenfeld, and both of them arrived in Israel via Europe in the previously mentioned Norsemans. They were flown out of Sde-Dov in the leased PAAC Dakota, piloted by Duval, to Geneva via Cyprus and Rome, and hence by train to Zurich and to Praga in a CSA airliner. The conversion course was organized by the Czechoslovakian AF at its base at Czeska Boduvitsa; most of the instructors had served with the Czechoslovak squadrons in the RAF during WWII. The Avia S-199 was a Messerschmitt Bf 109G fitted with a Junkers Jumo 211F engine instead of the original Daimler Benz DB605 because of the shortages in the supply of the latter. The Jumo 211 weighed more than the DB605, which produced 1475hp against the Jumo's 1350hp. Performance of the S-199 in comparison to the Bf 109G deteriorated considerably because of the center of gravity movement, and the Jumo's propeller that was larger and tended to roll the S-199 to the left on take-off, but it was a fighter, and nothing else mattered. In order to familiarize with German WWII vintage equipment, the pilot got some hours on Arado Ar 96B advanced trainers (also manufactured in Czechoslovakia as the Avia C2B1). For some of the pilots it was a "derusting" practice, but to others that had never flown fast and powerful aircraft, it was a first experience. It was soon revealed that the thought that such pilots could be converted to fighters in such a short conversion had no base in reality, and they did not finish the conversion stage. The pilots that did convert—Allon, Weizman, Cohen, Lennart and Rubenfeld—were all ex-WWII fighter pilots, and they first flew the CS-199, a dual seat operational trainer, and later on the S-199. Lennart made his first solo flight in an S-199 on 15 May.

Not a single shot was fired in that conversion, and when the Czechoslovakian instructors asked the Israelis, "what about to shoot" the eager pilots' answer was, "we will practice on the real targets!" And this they did effectively from 29 May onwards. The pilots graduated from conversion, and their S-199s arrived at Israel in an airbridge that was named Operation Ballak, after a biblical figure. Retroactively, the Hassida flight of late March was numbered Ballak 1, while on May 12 and 14, Ballak 2 and 3 arrived respectively. Ballak 2 was a leased Douglas C-54 Skymaster carrying 100 Vickers machine guns, but the time, in order to evade the American Embassy officials in Prague, the departing base was at Zatec, to be known by the mostly-American crews operating the airbridge as Zebra base. Ballak 3 was a leased CSA Douglas Dakota. Even earlier, on 3 May, the Curtiss C-46 Commando NX 67923 arrived at the ex-RAF station of Tel Nof from Italy and commenced supply flights within Israel. In one of those missions, this time to the besieged kibbutz Yechiam in the western Gallilee, it suffered ground fire damage and was grounded for repairs at Tel Nof.

The Sherut Avir was by now six months old. It was founded out of necessity by farsighted men with six aircraft and a few dozen dedicated members in November 1947. By May 1948 it had hundreds of men and its aircraft inventory had changed dramatically: it had three Lockheed Constellations, nine Curtiss Commandos, ten Avia S-199s, 19 Noorduyn Norsemans and many other types of lesser quantity and importance. Some of those aircraft, like two of the three Constellations, were never to see service with Chel Ha'avir, the successor of Sherut Avir, and most arrived after the first critical stage of the War of Independence, as Israel's struggle to survive

Home-made bombs. The shells were from cooking gas bottles.(IAF)

was to be known. On 11 May Aharon Remez reported to David Ben Gurion about SA's "order of battle": serviceable aircraft were three Austers, two Beech Bonanzas, one Dragon Rapide and one Douglas Dakota, while unserviceable aircraft included nine Austers, one Norseman, two RWD 13s, one Dragon rapide and one Curtiss Commando. The two Tiger Moths were also enlisted, but with the remark that both were written off. That total did not include the inventory of the Gallilee Sqn., two/three Austers and a single Fairchild, or the Negev Sqn. with two/three Austers as its sole strength. A number of other types, such as the Taylorcrafts and Seabees, were also ignored, possibly because they were not used for operational missions. The report clearly showed the poor serviceability rates of the period, but as mentioned before, fresh blood was on the way to Israel thanks to a motley band of colorful people—the purchasing agents.

Chapter 2: Buying Airplanes

An air force without combat aircraft is probably as ridiculous as a forest without trees, but that was the Sherut Avir situation up to the end of 1948. As already mentioned, in November 1947, when the Haggana's staff accepted in general lines the Remez/Shamir plan, it was decided to exploit the procurement possibilities of the world's largest free market, the United States, as the scene in which the purchasing experts of the Haggana would focus. It was soon discovered that the American market, although looking very promising, could not fulfill the initial hopes mainly because of the U.S. embargo of arms to the Middle East, announced on 5 December 1947, that was carefully imposed by the U.S. State Department. As a result, the attention of the Haggana's procurement efforts was turned in early 1948 to other possible sources, such as Latin America, South Africa and Europe. Most surprising of all, the nation that was to supply Chel Ha'avir with its first fighters, 25 Avia S-199s, was behind from what was to be known as the "Iron Curtain." After the S-199 deal, Czechoslovakia supplied to the Chel Ha'avir 50 Supermarine Spitfires, and although local air superiority was gained with the S-199 over central Israel, it was the beloved Spitfire that enabled the Israeli Air Force to gain air superiority all over Israel towards the end of 1948. But an air arm could not exist only with fighters. Transports, bombers, attack, observation and training aircraft were among the machines required to fulfill all the tasked missions. Those aircraft were bought and smuggled to Israel from all over the world and under the most peculiar covers the human mind can think of in such a moment of urgency: Boeing B-17 Flying Fortresses; Curtiss C-46 Commandos; Lockheed C-69 Constellations; North American P-51D Mustangs and North American T-6 Texans and Piper Cubs from the U.S.; De Havilland Mosquitoes, Bristol Beaufighters and a variety of light aircraft from the UK; and Noorduyn Norseman and Douglas C-47s from Europe, as well as many other types from Australia, South Africa, and elsewhere.

One of the first airplanes aquired by undercover Israeli purchasing agents abroad were ex-USAAF Norduyn UC-64 Norsemans, stored at depots in Europe.(IAF)

The story began in November 1947 when Yehuda Arazi was instructed to buy aircraft in the U.S. Arazi's thoughts were at first to purchase means to deliver the arms to Israel upon completion of the British withdrawal rather than buying the arms. All activity had to be completely legal, under carefully selected covers, and with very carefully selected American volunteers, most of them American Jews with considerable WWII flying experience. The main figures that helped turn Arazi's dreams and plans into reality were Al Schwimmer, Sam Lewis and Leo Gardner. Schwimmer was an aeronautical engineer who had served with the USAAF during WWII and was a flight engineer with TWA during the post-war years. Later on he became the Chel Ha'avir chief engineer with the rank of Aluf Mishne (Colonel), and was Israel Aircraft Industries' (IAI) general director for more than twenty years, from its foundation as Bedek Aircraft and up to 1977. Gardner was a civilian flying instructor before WWII and served in this role with the USAAF during the war years. Lewis was a very experienced pilot who learned to fly in 1933, became a civilian instructor two years later, and from 1938 was an instructor in the USAAF; in 1941 he joined the RAF as a ferry pilot, performing several flights over the Atlantic. A year later Lewis joined TWA as a Boeing 247 captain, converting to Douglas DC-4s two years later. Appointed as El Al's chief pilot, he retired in 1972 with 28.000 flight hours to his credit. All three of them approached the Jewish institutes in the U.S. during the summer of 1947 and proposed to buy a fleet of ex-WWII USAAF surplus transports to operate an airline that would relocate the Holocaust's survivors from camps in Europe to Israel, their future homeland. Initially they feared that their plan was ignored, but finally Schwimmer met Yehuda Arazi in New York and the scheme gained momentum. Arazi instructed Schwimmer to buy the aircraft, but made it clear that those aircraft were intended to transport arms to the hard-pressed Jewish community in Palestine—otherwise the Holocaust survivors would have nowhere to go to.

Schwimmer bought from the War Assets Administration five Curtiss C-46 Commandos for $5.000 each and three Lockheed C-69 Constellations for $15.000 each—the money was raised by Arazi from local American Zionists. The aircraft were all of WWII vintage and were ex-USAAF that had been in open storage for quite some time. Moreover, they were all to military standards and had to be civilianized and inspected by the FAA for their registration. Thus it was decided to form a refurbishment company for the operation, and Schwimmer Air Service came into being late in 1947. The first step was to lease room in the Burbank, California, airport maintenance area, and the second was to transfer the aircraft to the refurbishment center. For the Constellations it was an easy task, since they were parked in the same airport, but the Commandos were in Ontario another section of California, and considering their open storage for several months and the fact that the pilots that had to ferry them to Burbank either never flew a Commando in their lives or did not fly an aircraft since the end of WWII, the task was quite hazardous. But time was only one of the factors that was running out, and similar to many instances in the near future, risks had to be taken, and the Commandos were flown with their undercarriage down by Leo Gardner, Sam Lewis and three more volunteers; one of them was Bill Garson, ex-USAAF transport pilot with the rank of Major, who was killed a few months later when more risks were taken in order to ferry those same commandos to Israel. The creation of Schwimmer Air Service enabled the recruiting of more volunteers, both for air and ground crews, although most mechanics were hired without knowing the real destination of the aircraft, and spare parts and engines were bought to enable the operation of the fleet overseas once the refurbishment would end and the smuggling problem was solved.

Schwimmer Air Service was a good cover for the refurbishment of the aircraft, and although the FBI investigated their activities it seemed that the operation was advancing according to the laid plan. However, one link in the chain was missing, and that was a cover for the export of the aircraft and their operation overseas at least up to the creation of the state of Israel. Arazi searched for an airline that had ceased operations for some reason or another, but that was still registered as a completely legal company, and he found what he wanted in Service Airways. This airline was constituted by Irwin Shindler in 1944 as a small unscheduled airline in New Jersey, but had to close it down and was working as a copilot in another airline when approached by Arazi, who offered to recreate Service Airways to operate a fleet of three Constellations and five Commandos. Shindel, an ex-WWII USAAF transport pilot, agreed immediately, and was tasked to search for work for the reborn airline from new offices at 57th Street, New York City. Shindler was appointed Service Airways president with Steve Shwartz, an ex-USAAF navigator, and Ray Salck, another ex-USAAF man that previously worked with Schwimmer at Burbank as vice president. It was Shindler that discovered the almost perfect solution to Arazi's urgent problem, the export of the aircraft, by mere coincidence. He knew Martin B. Belfond, an ex-USAAF Major, that had founded after the end of WWII in New Jersey an airline with a single Douglas C-47. Belfond negotiated with the Panamanian government and was granted a unique opportunity to establish a national airline named Líneas Aéreas del Panamá, S.A. (LAPSA). The license granted Belfond with the right to operate Panamian registered airliners on worldwide routes and the Panamian embassies were instructed to help LAPSA gain landing rights all over the globe. Moreover, the license terms permitted the establishment of a refurbishment center in Panamá. It was a perfect cover that could enable the export of the aircraft and their operations over the Atlantic Ocean and in Europe, and it was the chance that Arazi had sought.

The export of the aircraft required major refurbishment and the deletion of all military equipment, such as communication sets, IFF transponders, machine guns and mountings, paratrooping equipment, etc. Gaining civilian registration meant another expensive and lengthy treatment that included such items as the replacement of the powerplant's military standard carburetors with civilian standard fuel injection systems. The civilian standard refurbishment of a Constellation could have a cost in excess of half a million dollars—more than thirty times the purchasing cost!. The full civilian standard refurbishment of a Commando was not possible at all at the time, and if one takes into account the time required for such an in-depth refurbishment and the time for FAA officials to inspect and re-inspect, then the Panamian registration was an excellent solution.

In January 1948 Yehuda Arazi was called back to Palestine to report his activities, and to be briefed about the latest developments. Arazi heard about the seven landing strips erected to receive his aircraft carrying arms from Czechoslovakia and was ordered to locate a base in Europe that would make such operations feasible. Angelo Ambrosini was the owner of a small aviation company at Castilliona di Lago in northern Italy and his factory suffered from worker's strikes and labor stoppages—not an unusual situation in postwar Italy. Arazi approached the Italian millionaire and simply told him the truth: a cover base was required for aircraft transporting arms to the Jewish community in Palestine. Ambrosini agreed to help, and the cover was that the intended LAPSA airliners would have their final furnishing at Castilliona di Lago before commencing operations as the "Panamian flag carrier"

But Arazi was also instructed to buy military types; the preferred aircraft were the Republic P-47 Thunderbolt and North American P-51 Mustang fighters, the Douglas A-20 Boston and North

The C-46 Commando was one of the most important machines acquired by the Israelis in 1948, serving as a logistic and tatical transport as well as a bomber. (IAF)

This Republic Seabee amphibian served as a hack with 101 Squadron. (IAF)

American B-25 Mitchell light/medium bombers, the Boeing B-17 Flying Fortress heavy bombers, and the North American AT-6 Harvard trainer. All were available, at bargain prices, in the U.S. and also in Latin America, but the problem was to export them. Arazi was given ten North American B-25 Mitchells, ten Grumman F6F Hellcats, five Lockheed PV1 Venturas and a pair of Grumman TBF Avengers freely from a group of supporting Jews in Hawaii, but none of these ever arrived in Israel due to the high risk involved in smuggling them. It was also during January that the first refurbished Lockheed C-69 Constellation took off from Burbank and set an easterly course to Melville, New Jersey, the base of Service Airways and also where the Customs Department inspection was to take place before the actual export. The plan was to concentrate the refurbished aircraft at Melville under a restricted FAA type license and proceed from there to Tucoman Field near Panama City. The crew of the first Constellation to cross the U.S. on 26 January 1948 was Sam Lewis, Captain; Leo Gardner, Copilot; and Al Schwimmer, flight engineer. Al returned immediately to Burbank to organize the movement of the workshops to Panama. The final contract between Service Airways and LAPSA was signed on 15 February 1948, and a formal request for a temporary export license for refurbishment abroad of one Commando and one Constellation was forwarded to the Department of State. The first refurbished Commando (NX67923) was flown to Melville on 28 February, and the imaginary scheme incepted less than half a year before was beginning to turn into a reality.

While Yehuda Arazi operated mainly on the American continent during that period, another aviation procurement agent was working in Europe. Harry Thorsh Fredkins, or Freddy Fredkins as he was known, was an ex-RAF Wellington pilot that settled, after his demobilization from the RAF, in the Belgian Congo and became a businessman. However, Fredkins had close ties with the Haggana movement, and in late 1947 he returned to Palestine and asked to join the Sherut Avir as a pilot. Instead he was sent to Europe to buy aircraft. Unlike Arazi, Fredkins' budget was much lower and his plans were less ambitious than those of his colleague.

Fredkins concentrated on the procurement of light aircraft that were to be legally transferred to Palestine as "Aviron's Aircraft", and formal import licenses were asked from the British Department of Civil Aviation in Palestine, headed at the time by one Mr. Storer, who promised to grant import licenses for one De Havilland Dragon Rapide, four Avro Ansons, six De Havilland Tiger Moths and a single Percival Proctor. During December 1947 Fredkins negotiated in England the purchase of eight De Havilland Tiger Moths, two Percival Proctors, one Auster and three Avro Ansons. Finally, he purchased six Ansons, one Dragon Rapide, six Tiger Moths and a single Proctor. Fredkins recruited aircrews in England for the ferrying of his aircraft, and one of them was Dan Tolkowski, an ex-RAF WWII pilot who was studying engineering in the U.K. at the time and years later was to become the Chel Ha'avir fourth commander. Nevertheless, Fredkins experienced severe difficulties in his attempts to deliver the aircraft to Palestine, and for that reason it was decided during February 1948 to send more procurement agents abroad. On 1 February Iseek Ashel, Sherut Avir's commander, approached Ben Gurion and presented him with a plan to procure three Douglas C-47 Dakota transports, ten North American B-25 Mitchells and three Avro Ansons. All were to be based in Italy until the British withdrawal was completed and then flown to Israel. Ashel also asked to send Aharon Remez and Heyman Shamir to the U.S. for procurement of aircraft and recruiting of trained air and ground crews. It seems that the SA's men were not over-enthusiastic about Arazi's Constellations and Commandos, and they probably preferred smaller and more versatile machines for their daily operations within Palestine. Another procurement agent that has not been mentioned yet was Boris Senior, an ex-SAAF fighter pilot and one of the founders of the Tel Aviv Sqn., who was relieved of his position and was sent to his birthland, South Africa, to purchase aircraft and recruit volunteers for the Sherut Avir.

Senior reached South Africa on 15 February and immediately made contact with a group of Jewish businessmen that aided him in the procurement of 50 Curtiss Kittyhawks that were sold as scrap at Kimberly for 300 Sterling Pounds!. The enthusiastic Senior, although realizing that fighters such as those were indeed the machines required, soon understood that there was no possible way to ship the Kittyhawks to Palestine and commenced searching for civilian aircraft that could be smuggled out of South Africa and still give valuable service to the Sherut Avir.

Senior purchased more than ten civilian aircraft and set to plan their ferrying to Palestine across the black continent. Even nowadays no one can consider a flight of a civilian light aircraft from South Africa to Europe or the Middle East as an easy task, but to Senior and his fellow pilots that he recruited, it seemed the only solution. The aircraft were to be ferried in singles by their "young rich owners" on a "pleasure" journey to Europe, while the transports were to be registered by real or fictitious airlines and also ferried to the north. The routes were to be carefully planned: flying through east Africa was shorter, but necessitated overflying through hostile Arab nations, while the west Africa route was not only longer but also necessitated overflying British territory or British influence countries, and both were very unfavorable. Because each pilot planned his own route and was completely on his own the moment he left South African territory, it took each a variable amount of time to reach Palestine. In fact, the aircraft which had the most uneventful flight of all, although the slowest in terms of time, never arrived in Palestine!.

The first of three Fairchild Arguses, flown by T. Blair, landed at Sde-Dov on 15 May 1948, the first day of the state of Israel. The other two Fairchilds arrived shortly afterwards and at least one source claims that they were PAAC registered and flown through Cairo, although it may be unlikely considering that they all arrived at Israel after 15 May, when Egypt was already deeply involved in an all-out war against Israel and also considering the fate of another aircraft purchased by Senior.

Senior procured two De Havilland DH-89A Dragon Rapides, registered by PAAC and ferried separately. The first arrived, but the second, G-AKOI (c/n 6546), was interned by the Egyptians that claimed that PAAC was maintaining an air link with Israel, which was in a state of war with them. The Egyptians released the foreign crew, but sold the aircraft to a Jordanian company named Arab Airways Association, and it was then registered as TJ-AAQ in July 1948. However, the story of this unfortunate Dragon Rapide did not end there, and we shall meet it again on a later chapter. A lonely Avro Anson was also purchased by Senior, but its fate is not clear: some sources claim that it crashed on its way to Israel, while others insist that it reached Sde-Dov after several weeks of journey via west Africa. Senior purchased two Douglas C-47s (ZS-AVM and ZS-AVO) that reached Israel under the cover of a fictitious airline named Westair during the early summer of 1948 and leased two other Dakotas, one from PAAC and the other from Universal, for their air link between Israel and Europe as already mentioned. The PAAC Dakota (ZS-AVK) reached Palestine in April 1948 and the Universal Dakota (ZS-DCZ) a bit later. The Dakotas were leased by Westair for a "proposed" charter line between Johannesburg and Tel Aviv, through Rhodesia, Uganda, Kenya and Sudan, that obviously never materialized, although the Dakotas found their way to Tel Aviv. By far the most interesting part of Senior's procurement trip to South Africa was the Beechcraft Bonanzas story; this light aircraft had been introduced in the market only a year earlier and was considered one of the top light-planes of those days: it was fast, well equipped, and had superior performance compared to the assortment of light aircraft then operated by the Sherut Avir and its sectorial squadrons. Senior bought two Bonanzas (ZS-BTE and ZS-BWS) and, after making all the arrangements necessary for the ferrying of the rest of the aircraft he purchased or leased, he decided to fly one of the Bonanzas to Palestine through east Africa accompanied by the second Bonanza flown by a South African volunteer named Cyril Katz. Their cover was a luxury tour of Europe by two young rich men. They left South Africa in April 1948 but went no further than Rhodesia. After an uneventful flight to Pitsbain they headed to Losaka, but only Senior arrived. Katz had to perform an emergency landing on a sand bank in the Zambezy river and slightly damaged his Bonanza, and when the worried Senior was informed by the local police about the incident he immediately flew to the emergency landing location with a mechanic, only to damage his Bonanza on the same muddy sand of the river bank. Finally they managed to repair one Bonanza (ZS-BWS), and Senior returned to South Africa while Katz was left to take care of repairing the second Bonanza. It took Senior several weeks to recover from the Malaria that he caught in his first attempt, but this time the South African authorities, well aware by now of his activities, informed him that he could fly wherever he wished within South Africa, but should not try to cross the borders with his "private" Bonanza. The trick adopted by Senior was a very elementary one, but it worked quite well; an ex-SAAF pilot named Cecil Wolfson was hired to fly the Bonanza to Wadi Halfa on the Egyptian-Sudanese border. In order to remove any suspicious Wolfon's wife accompanied him for the "pleasure" tour while Boris Senior was to board a regular airliner to Wadi Halfa and continue from there to Palestine with the red and silver painted Beechcraft. The plan was precisely executed, and Senior collected the Bonanza at Waddi Halfa and flew to Luxor in the Nile Valley in Egypt where he spent the night in a hotel without anybody suspecting that the "rich" young South African was smuggling an aircraft. Early the next morning Senior filled a flight plan to Beirut but "aborted" to the Negev Sqn. base at Nur Am. He

The Douglas C-47 proved, as in other parts of the world, to be a real workhorse, serving the Israelis from their earlier days until the present times. (IAF)

immediately proceeded to Sde-Dov where once more he took the air base commander post and his Bonanza was immediately sent to an observation flight of the blocked Tel-Aviv/Jerusalem road. It was in this second week of May–before May 11–that the other Bonanza, flown by Cyril Katz, arrived. Yet another Bonanza arrived at a later date, but its origins are not clear. At least one source claims that three were purchased by Senior in South Africa and sent in crates to Italy where they were prepared for delivery to Israel, but it may well be that the Bonanza concerned was bought in Europe by another procurement agent, perhaps Fredkins. The only reliable witness concerning these aircraft was the pilot who ferried one of them to Israel. Wim Van Leer was a young Jewish millionaire from the Netherlands whose father donated a Dakota to the Chel Ha'avir, as well as two Grumman G44 Widgeons on a later date. Van Leer claims that he was ordered to ferry the Bonanza to Israel and insisted on planning his own route; he probably commenced the ferry flight in Italy, but his first stop was Amsterdam, where long range fuel tanks were installed in the KLM workshops. He then flew along the North African coast through Bengazi and the RAF station at El Adem, Lybia, and from there he continued directly to Haifa.

Two other men, Avraham Rotenberg and Chai Issaschar, both connected to the aviation world, left Palestine for London during February. Rotenberg was the director of the Palestinian electricity company that his brother, Pinchas Rotenberg, founded. Pinchas also enterprised the creation in the mid-thirties of Palestine Airways, already mentioned. Issaschar was a senior engineer in the electricity company and a senior Haggana member. As early as January 1948, the problem of the air link between the Jewish community in Palestine and the outside world arose. Rotemberg wanted to organize a civil service with the cooperation of a British company named Sky Ways, and the planned routes were: Lod-Paris, Lod-New York, Lod-Athens-Rome, Lod-Cyrprus and Lod-Haifa. Nothing came out of that plan, and in February Rotenberg and Issaschar left Palestine for London to purchase three Douglas Dakotas for such a civil service. Their mission failed, as the British Foreign Office objected the granting of import licenses for civilian aircraft, although the Palestine Governor and the Colonies Office had approved the requests. Henceforth, no civil aircraft could be legally imported to Palestine, and only when Senior's Dakota became available from April 1948 onwards was the air link to Europe set up.

The Foreign Office decision affected Fredkins, too. Fredkins was quite sure that his aircraft would encounter no problems on their way to Palestine. In February he ferried five Avro Ansons and a De Havilland Dragon Rapide to Paris, and a Percival Proctor to Marseille. During the following month those aircraft, and another Anson, were flown to Italy, while the six Tiger Moths were left behind in England. On April 10 four of the Ansons left Brindisi for Palestine under the cover of a ferry flight to Singapore, but were interned at Rhodes by the Greek authorities and British intelligence informed them that the Ansons were flown by "Communists." Since it was a period of a bloody civil war against the communists in Greece, the Greeks took the false information seriously, and the crew members were imprisoned for several weeks. One of them was Dan Tolkowski, and another was Yehuda Leib (Leonard) Cohen, a very experienced ex RAF WWII, who later would be killed in

In December 1948 this Grumman G-44 Widgeon amphibian crashed into Lake Kinneret (Sea of Galilee), killing its three crewmembers. (IAF)

Rome on May 20 while checking out the 32-victory Canadian ace George "Buzz" Buerling on a Noorduyn Norseman. After several weeks in the local jail the pilots were finally released and returned to Italy, but the Ansons were not and they joined IDFAF service a year later.

The Ansons failure was a major blow, but Fredkins was already deeply involved with his largest aircraft deal. While in Paris he contacted David Miller, an ex-USAAF Lt Colonel who, after WWII, became an aircraft dealer in Paris. Fredkins told him that he was the representative of a Belgian company named Somaco planning a cargo service to Spain and asked Miller if he could propose any suitable aircraft. Miller offered the Noorduyn Norseman and told Fredkins that the USAF was selling 50 UC-64As, as the Norseman was known, at bargain price. Fredkins was authorized to buy 20 in early April after he made it clear that five could be delivered immediately, another seven up to May and the remaininder henceforth. The Norsemans were flown from the USAF base at Oberpaffenhofen, near Munich, to Amsterdam, where they were inspected in the KLM workshops and long range ferry tanks installed in the cabin. From Amsterdam the refurbished Norsemans were flown to Toussus-le-Noble, near Paris, where the crews waited to collect the aircraft and ferry them across the Mediterranean to Palestine. The ferry route was surveyed by Wim Van Leer, who tried to locate possible emergency airfields. The first Norsemans were ready by late April, and two were ferried to Palestine, via Italy, in the first week of May. The arrival at Sde-Dov of such a "big and powerful" aircraft was welcomed by the Tel-Aviv Sqn. pilots, used to Auster and the like in those days, although the Norseman was never liked by its crews, it was sluggish in Israel's hot climate, especially when overloaded as was the usual case, and its narrow track landing gear made ground loops a familiar habit. The ferrying of the Norsemans lasted up to the end of the summer of 1948 due to a slow refurbishment rate and other reasons to be discussed later. Milton Robenfeld, Lou Lennart and Coleman Goldstein, Jewish volunteers from the U.S., were among the crews that delivered the first two to Sde-Dov, while the rest of the crews were mainly the American pilots, recruited by Heyman Shamir in the U.S., which reached Europe with regular commercial airlines and concentrated in Rome, from where they were assigned to the ferrying missions.

Heyman Shamir was the last agent. He was sent in February 1948 and he had to recruit air and ground crews and coordinate the purchase of bombers and fighters in the U.S. Until Shamir's arrival a small number of aircrews were recruited by Schwimmer, Gardner and Lewis, but as needs grew and so did the number and complexity of aircraft acquired, it was obvious that much more experienced pilots, especially of fighters and multi-engined heavy transports and bombers, had to be recruited. The first steps were to locate Jewish pilots with WWII flying experience. Most of the pilots approached volunteered, and others did not wait for the agents to knock on their doors and found the right persons using the established Zionist institutes for the first contact. Later on many American non-Jewish pilots volunteered either because of the adventure and risk that was missing since WWII had ended or because of ideological reasons, namely to help the people of the Bible to establish their home in the Holy Land.

While Shamir assumed responsibility for the recruiting of air and ground crews, Schwimmer and his friends were able to concentrate on refurbishment and ferrying the Commandos and Constellations. The temporary export license for one Constellation and one Commando was granted in early March, and after a Customs Department clearance check at Teterborough, New Jersey, on 6 March the Service Airways marked Commando NX67923 commenced its long flight to Italy through Goose Bay, Blue Weist 1, Iceland, Shannon and Geneva. The crew was composed of three pilots, Leo Gardner was the captain, and the others were Arnold Ilwhite and Robert Fine, with three other crew members: Irwin Shindler was the radio operator, and Arnold Stahalic and Sam Pomerance mechanics. The flight was uneventful except for an unfortunate incident that turned the international press attention towards Service Airways and LAPSA, a most unwelcome matter for a clandestine operation. The C-46A arrived to Ambrosini's airfield at Castilliona di Lago as planned on 9 March, but the airfield lacked any Italian authority representatives, the crew did not bother to inform about their arrival, and the press immediately published that a Service Airways Curtiss C-46A Commando was missing between Geneva and Italy. The publication of the story led to many reporters trying to investigate Service Airways' background and its connections with LAPSA, and the worried Schiwimmer not only had to try to get first hand information as to the fate of the missing Commando, but also had to tackle the reporters.

The mystery was finally solved when Arnold Ilwhite phoned his father in the U.S. to inform him that he was just fine, although the damage to the secrecy of the operation was already done. Moreover, the Castilliona di Lago runway was found to be too short and too soft for such a heavy transport and that convenient base had to be abandoned; the Commando was flown to nearby Perugia airport and stayed there for the next two months. The dilemma at the time was whether to use the Commando to transport the arms bought in Czechoslovakia to Palestine, where it was desperately required, or to use the scarcely viable leased transports. The Commando was indeed available, but a regular link between Prague and Palestine

The Boeing-Stearman PT-17 Kaydet was the IDFAF's main primary trainer from the late 1940s into the mid 1950s. (IAF)

The Douglas C-54 was used by the Panamanian-registered LATA airlines on airbridge duties mainly from Europe to Israel. One of them was the first aircraft in El-Al colors and carried Israel's first President, Mr Chaim Weizman, to his country.

could not be hidden from the Americans, and although the Cold War was not yet at its peak, the Iron Curtain sensitivity was a major consideration in the policy of the Department of State and it was anticipated that such an activity would unveil the true story behind Service Airways and LAPSA with consequent results. Since most of the aircraft were still at the U.S. awaiting refurbishment and export licenses, the obvious and reasonable decision was not to risk what had been gained so far and that Commando had to wait for more than two months before crossing the Iron Curtain for the first time. This aircraft was not used to deliver arms to Palestine at that stage, but for denying them to the Syrians. The story of the vessel Lino search was told in Part 1 of this narrative, but it seems that the decision not to use the Commando on flights to Prague in order not to risk the whole operation was logical, though the decision to use the same aircraft to attack a small Italian ship in international waters with an improvised 20mm gun and home-made bombs was not quite rational and the pilots did question the logic of the scheme. Nevertheless, they did fly each day and among them were also Wim Van Leer and Dan Tolkowski, previously mentioned in conjunction with Fredkins' operations. The time was used to fit long range fuel tanks in the Commando for its anticipated role of ferrying arms from Czechoslovakia to Israel, and during April Gardner and Shindler returned to the U.S. to participate in the ferrying of the other C-46 Commandos.

Robert Fine was instructed to join the Norseman ferrying operation, Arnold Stahalic was sent to Amsterdam to inspect the Norseman refurbishing work, Sam Pomerantz was sent to Czechoslovakia for the acceptance checks of the S-199s and Arnold Ilwhite was left in charge of the Commando until it was the first of its breed to reach Palestine on 3 May 1948.

The second aircraft that was granted an export license was the Constellation, already at Malville since January, and it too passed the Customs Department's procedure at New Jersey on 13 March, already with the markings of LAPSA and carrying the Panamian registration RX-121, with Sam Lewis flying it to Panama via Jamaica. The Constellation was greeted at Panama on a formal ceremony as the flagship of the new national airline, LAPSA, was being leased to one of the candidates for the local elections as well as to the local Pepsi Cola dealer, while Sam Lewis used the spare time to give familiarization flights to other pilots and to recondition the Constellation with the aid of some mechanics who were flown to Panama in the Constellation with Lewis.

Soon afterwards this aircraft was grounded due to engine troubles and a replacement powerplant was sought, an uneasy task considering the recent developments. In the fourth week of March 1948 President Harry S. Truman announced that, commencing from midnight 14 April, export of civil aircraft, spare parts, avionics and radars would have to be approved by the State Department in the same procedure that already applied to military aircraft. The publicity of the Commando incident turned the attention of the FBI and State Department to Service Airways and LAPSA, and the new regulation put the operation at stake unless immediate action was taken. Even before President Truman announced his order, five more Curtiss C-46As were procured from the War Assets Administration, while a pair of partly-refurbished Commandos were at Melville awaiting final repairs before departure to Panama. It was decided to complete the semi-refurbishment of the Commandos still in California and fly all of them to Panama before the resolution became valid on 15 April. It was a magnificent effort, and on 9 April four Curtiss C-46A Commandos were ready to leave Melville to Panama. As it was obvious that it was a one way trip, the aircraft were loaded with the equipment considered essential to operate them abroad with minimum maintenance—spare engines and parts and enough to even complete Vultee BT-13 trainers, out of five that Arazi acquired in late 1947 and which were flown from Melville from Boston by Norman Monitz and Larry Rabb, two of the pilots recruited for the big adventure of flying half the globe from the U.S. to the Middle East through South America. The Commandos were painted in LAPSA colors and carried Panamian registrations, but it seems that the American authorities knew exactly their whereabouts and the Customs Department tried to suspend their departure knowing that only five days were left before the new regulation would prohibit the export. Al Auerbach, an ex-WWII U.S. Navy Lt Cdr who quit his job as an FAA inspector to lead the four Commandos out of the U.S., handled the negotiations with the Customs Department, and on the evening of 10 April four veteran C-46As and their determined crews took off from Melville. The plan was to fly separately to Panama via Jamaica, but only one Commando, flown by Service Airways president Irwin Shindler with William Garson as copilot, corresponded with the original plan. The rest made refueling stops at other locations: Norman Monitz at Nassau; Larry Rabb at Cuba; and Auerbach at West Point Beach, Florida. Aluf them arrived at Tucoman on 11 April. The first long range overwater flight was completed successfully, and now the next stage, a long range flight over land, had to be executed. Auerbach, Shindler and Garson left Panama for California to help with the second stage of Schwimmer's aircraft big departure. The five remaining Commandos were hurriedly readied at Burbank to escape the 14 April midnight deadline, and the workshp equipment was stuffed into the overloaded aircraft. On the noon hours of 14 April, the five machines, led by LAPSA's chief pilot, Sam Lewis, took off from Burbank to fly along the Pacific Coast to Panama, although once more the plan was not executed as briefed, only this time it cost the lives of two volunteers. Instead of flying following the Mexican coastline, the aircraft stopped at Mexico City, ignoring their overloaded condition and the poor performance of such a heavy-loaded aircraft at a high-altitude take-off. Adding another mistake, they took off at noon on a very hot day, and it was the destiny of the third Commando to crash. William Garson, the sole pilot aboard, and Ernest Glen King, a mechanic, were killed. The two Commandos that took off landed, and the overload, estimated at thousands of pounds per aircraft, was unloaded and stored at local warehouses for later collection. The next day, with a deteriorating morale, the remaining four aircraft flew uneventfully to Panama, but the unrest of the air and ground crews was such that Irwin Shindler, probably considered responsible for the overloading, was replaced as LAPSA president by Sam Lewis, and a small number of volunteers, air and ground crews, decided to leave the operation and return to the US. While all Commandos left the U.S. under Department of Commerce approvals for temporary stay abroad for refurbishment, the two Constellations were left behind since they weighed more than 60,000 Pounds and thus required the State Department's approval as well as that of the Commerce Department (these approvals were valid up to 14 April midnight). The final answer of the State Department to the formal export requests, given on 16 April, was not only negative, but also demanded that the third Constellation RX-121 should be returned to Panama!. The period spent in Panama was exploited for training the crews and repairing the aircraft. It was found out that most of the aircrews lacked Commando experience, and a general familiarization—as well as navigation and instruments flying training—scheme was commenced while the poor technical state of the aircraft was taken care of by the ground crews towards the anticipated Atlantic Ocean crossing. But transports could not win a war, fighters were needed, and Yehuda Arazi was well aware of that fact. Fighters of all sorts could be bought in the U.S., but could not be exported. On the other hand, smuggling them in crates via the sea routes meant a long period of delivery and reconditioning to combat status, and time was the only thing that was really running out. Arazi turned his attention to Latin America and both he and Schwimmer found what they were looking for; 25 P-47D Thunderbolts in Mexico—not the favorite P-51 Mustangs, but still a most capable and valuable fighter. The deal cost about $1 million U.S.,

The peaceful Beechcraft Bonanzas acquired in South Africa by Boris Senior were quickly converted into makeshift bombers by adding a rough coat of camouflage and some crude bomb racks. (IAF)

and the P-47s were to be flown within LAPSA aircraft from America to Israel, from where the experienced American volunteers would fly them. It was an excellent plan except for one detail: the Haggana representatives had already ordered fighters for the future state of Israel, and there was not any money left, or will to risk, for the Mexican P-47s. On 23 April the Haggana representatives at Czechoslovakia signed a contract for ten Avia S-199 fighters valued at 1,800,000 dollars—quite expensive, to say it mildly—the price was abusive, if we consider the fact that the S-199 was the old Bf-109 powered by He-111 engines since none of the original powerplants existed in Czechoslovakia. Each Avia cost 44,600 dollars, its associated equipment 6,890 dollars more, and the ammunition cost per S-199 was 120,229 dollars. The total cost of each S-199 was thus 171,719 dollars, which meant a contract value of 1,717,190—the balance was made up of ferrying fees. Schwimmer argued that most of the future combat pilots were American volunteers who would hate the idea of fighting in a German/Czech hybrid (known as the Mezec-Mule) and that the P-47 was far superior to the S-199 on all accounts, but the deal was signed and the Haggana leader preferred this time to take the safer side of the road. The contract included spare parts, technical assistance, and conversion courses for a small number of pilots who were supposed to collect the S-199s from the factory and ferry them to Israel via Italy and Greece. However, the arrest of the Ansons at Rhodes meant clearly that such a scheme was not feasible and that the fighters would have to be dismantled and ferried to Israel by some other way. A land and sea voyage was not considered because of the critical time factor; if the fighters would be late they might not have any soil to protect, and the obvious move was to bring Schwimmer's transports from Panama and establish an air bridge between Czechoslovakia and Israel. The difficulties were obvious, too: the Commando could not fly directly from Czechoslovakia to Israel (a distance of some 3.000 kilometers) when loaded, and a discrete refueling point had to be organized somewhere in Europe. As total air superiority was expected to be enjoyed by the Arab nations at least during the early stages of the conflict, the arrivals and departures should be performed under the cover of darkness and the transports could not spend the daylight in Israeli airfields due to the fear of air attack. The answer to the two last restraints was to establish the operations and servicing center at Czechoslovakia, where a complete airfield, Zatec—officially codenamed Etzion and nicknamed Zebra by the foreign crews—was occupied mainly by American citizens. The first problem was solved by using Ajaccio airfield at Corsica, France, as a refueling stop, although the difficulties with that point were not overcome until the airbridge ceased operations in August 1948. The distance from Czechoslovakia to Corsica is some 1.000 kilometers, and from Corsica to Israel is about 2.600 kilometers. Another point, which may be mentioned here as well, was that the fledging Chel Ha'avir's staff, hardly controlling the operations of the fighting units, could not handle efficiently such a large scale, regular based operation, and in late May 1948 Lahak Tovala Avirit (Air Transport Command), LATA, was established, commanded by Meir Mardor, a veteran Haggana commander and procurement agent without aeronautical background.

The creation of LATA and the transport of the S-199s to Israel was impossible without the Commandos, and in early May, after almost a month of refresher training and servicing, five Commandos were ready to fly east across the Atlantic Ocean to Europe. The Constellation and three more Commandos were left behind for more in-depth maintenance work under the command of Sam Lewis. Al Auerback led five Commandos loaded with the two Vultee BT-13s, leaving Tucoman on 8 May, and only two of them arrived in Europe on 15 May, the very first day of the state of Israel existence. They flew uneventfully via Paramaribu and Port of Spain to Natal, Brazil, where after two days of servicing it was decided to leave two unserviceable Commandos behind and proceed across the Atlantic Ocean to Dakar with the remaining three. The two Commandos left behind were RX-133, flown by Trig Messang and Al Reisin, and RX-136, flown by Norman Monitz and Sheldon Icle, while the three that took off during the early evening hours of 13 May were: RX-130, piloted by Al Auerbach and Bob Laury; RX-135, piloted by Martin Ribcoff and Ray Kortz; and RX-138, piloted by Larry Rabb and Philip Shield. RX-130 returned to Natal due to engine troubles, but Ribcoff and Rabb pressed on and landed at Dakar on the morning of 14 May. Time was running out, and the pair of old Commandos continued immediately to Casablanca, French Morocco, and Catania, Italy, where they were greeted by Daniel Agron, the Chel Ha'avir representative in Italy, and Heyman Shamir, who was to navigate them to Tel Nof air base, Israel. One great adventure was over—the other had just begun.

The UC-64 Norseman served the IDFAF in the liason, light transport and bomber roles during the War of Independence.

This curious picture taken at Sde Dov shows the only Aerovan which served for approximmtely a month with the Chel Ha'avir.(GPO)

Chapter 3: Against The Wall

In the early evening hours of Friday, 14 May 1948, David Ben Gurion, the leader of the Jewish community in Palestine and soon to become the first Prime Minister and Minister of Defense of Israel, announced the birth of the new state. The citizens of Israel forgot for a few hours the combats and troubles and danced in the city's squares and streets to celebrate the realization of a dream that might have been very short. Five established Arab nations challenged the fledging state and tried to ruin it from the start. Egypt, Syria, Lebanon, Jordan and Iraq committed their armies for the task. They were well equipped and well trained, and the only point in which the Haggana forces, the biggest Jewish movement in Palestine that formed the basis for the IDF (Israel Defense Forces), could claim any advantage was motivation. While the Jewish were highly motivated to handle their own fate, especially considering the horrible holocaust suffered by the Jewish nation in Europe during WWII in which six million Jews were murdered by the Nazis, the Arab armies were rather expeditionary forces fighting on a distant front to "aid" a brotherhood nation, the Palestinians, in their struggle to achieve the holyland and not to share it with the Israeli nation as the United Nations decided on 27 November 1947. Thus, the Arabs were fighting on a relatively distant frontier for what could be termed as another's cause, while the Israelis fought for their very real existence. Those Arab armies, in contrast to the Israeli forces, had uniforms, discipline, orders and all these other little things that make a regular army what it is. The Israelis lacked uniforms and unit construction, and orders were very elastic during the early days. That point might be considered by many as an advantage, but it must be considered that the IDF was composed of veteran Israelis, holocaust refugees from Europe, and Jewish and non-Jewish volunteers from all over the world. The situation was such that the first armored regiment organized by the IDF had two companies, but in each of that companies a different language was spoken, and neither of the two was Hebrew—Russian on the first and English on the second!. Moreover, while a regular army possesses standard weapons, with spare parts and workshops for maintenance, the Israelis were operating a large collection of weapons purchased all over the world or even assembled in Israel from parts found in ex-British Army depots. But an army cannot rely only on weapons—a large number of other items are on the shopping list: ammunition, communications equipment, food, clothes and more and more. Aluf that is handled by a large logistics department, which the Arabs had, while the IDF had to create it from zero, an uneasy task during

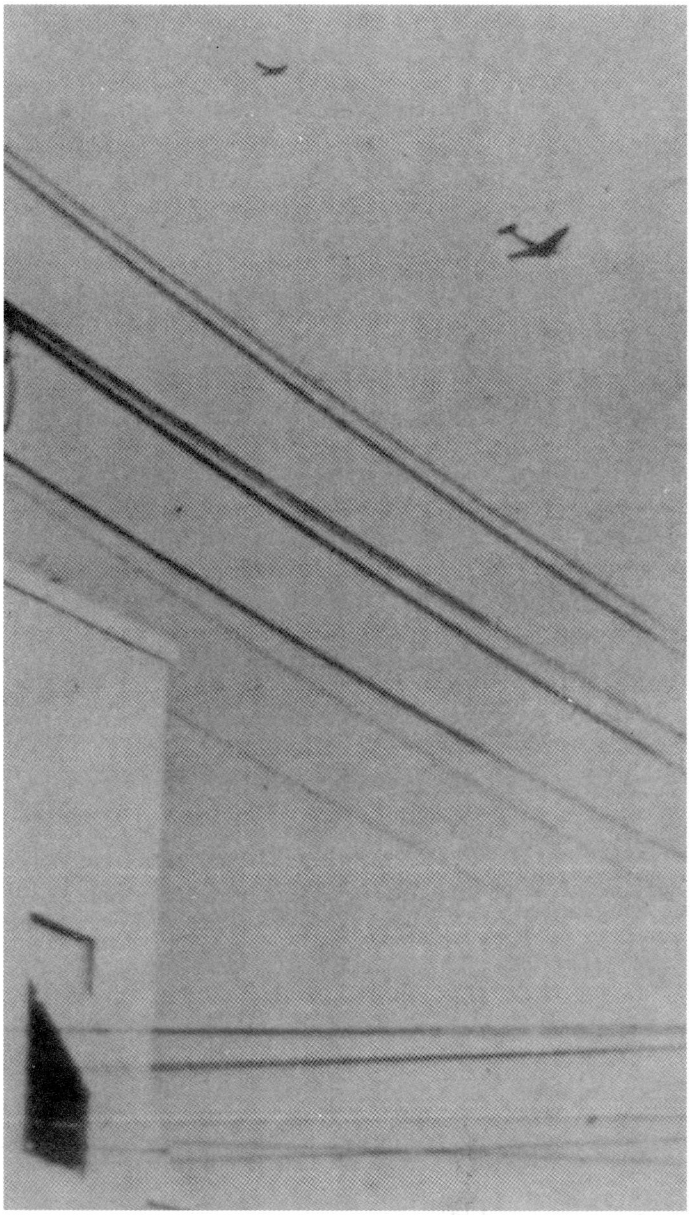

The Israeli Air Force's first of a long list of air-to-air kills during its history were two Royal Egyptian Air Force C-47 Dakotas which were bombing Tel-Aviv on 3 June 1948. The victor was Modi Allon, 101 Squadron first C.O.; the photo shows Allon closing to point blank range in his S199 for his second kill. (IAF)

a period of war. Difficulties not withstanding, the Israelis had nowhere to go, being surrounded by four hostile Arab nations in the north, east and south and by the Mediterranean Sea to the west. It was a war of survival, although it has been known ever since as the "War of Independence."

The Arab armies had tanks and artillery and ships, and most importantly for this narrative, all, or nearly all, possessed aircraft. The Royal Egyptian Air Force (REAF) was the best equipped of the Arab air arms and by then had existed for more than a decade. The REAF saw its origins in the early thirties as the Egyptian Army Air Force under the command of an RAF officer, Air Commodore A.G. Board, to signify the close ties between the fledging EAAF and the RAF in Egypt that mainly protected the Suez Canal Zone, which was under British control. British training and liaison aircraft were operated during the first period, and in 1939, when the title was changed to the REAF, the service had a Central Flying School at Almaza, near Cairo, and two operational squadrons: No 1 with Avro 6266s and No 2 with Hawker Audaxes. Following this retitling a major expansion scheme was adopted and the REAF took on strength with 42 Miles Magister trainers, 20 Westland Lysanders for army cooperation, and 18 Gloster Gladiator fighters into a new structure of a five squadron force: No 2 and No 5 with Gladiators, No 1 with Lysanders, No 3 with Avro 626s and No 4 with the remaining Audaxes. The Egyptian neutrality during WWII decreased the amount of cooperation with the RAF; nevertheless, a large number of quite modern ex-RAF aircraft entered service, including Supermarine Spitfire Mk.Vc, Hawker Hurricane Mk.IIB/Cs and Avro Ansons. By the end of WWII the expanding REAF had concentrated in bases at the Nile Delta, while the RAF was still occupying the Suez Canal Zone bases. The Spitfires were augmented by the improved Mk.IXs, while the Hurricanes were phased out of service. "Bombers" were added to the inventory in the shape of ex-USAAF Douglas C-47s converted for that role, as well as a small number of Short Stirlings, while heavy lift capacity was achieved by a small number of also ex-USAAF C-46 Commandos. Before the war's end the REAF would reinforce their depleted Spitfire ranks with the acquisition of Macchi MC.205Vs from Italy. In May 1948, when the REAF was called for its real operational debut, it was a well organized force with strong British affiliation and fine tradition. Moreover, it had several months for preparation and intelligence gathering—for example, an Egyptian aircraft on a flight from Lod to Cairo on 17 February 1948 passed above Tel-Aviv and photographed the harbor.

Although the REAF later suffered from the U.N.'s embargo on the Middle East warring nations, its status at that time could be termed as excellent when compared with other neighboring air arms of the period—dominating the battle area for the first stages of the conflict, though it later losing its air superiority, the REAF still continued operations to the very end of the War of Independence. But the REAF had its share of problems, too, the main problem being that the Egyptian ground forces were like an expeditionary force, the same applying to the REAF units which fought the conflict. The long distance between the REAF's main bases in the Nile Delta and the Negev, the area in which the Egyptian and Israeli forces clashed, was beyond the range capabilities of its aircraft, except for the makeshift Dakota bombers and later the Stirling, which had a very poor serviceability while in REAF service, and thus an air base near the front was required. That requirement was fulfilled by the El Arish air base in northeast Sinai peninsula, but that base was completely detached from the REAF main facilities and had to rely on its own services, maintenance, logistics and administration, with obvious difficulties arising from such a situation. A large number of modern and well equipped air bases existed in the middle, but all were RAF-manned in the Suez Canal Zone, stretching the communications and supply lines even more and complicating the situation. The main burden of the REAF's endeavor fell to the No 2 and 6 Squadrons equipped with Spitfire 9s at El Arish. These beautiful fighters were used for both air-to-air and air-to-ground missions. No 4 Squadron C-47s were briefly used as bombers, and later only as transports to supply the forward air bases

The second fighter to serve with 101 Squadron was the incomparable Spitfire. The first few were acquired from scrap dumps in Israel, the others acquired in Czechoslovakia. (IAF)

of El Arish and Bir Tmadda (to be constructed and turned operational towards the end of 1948) in Sinai. There are some sources that point out that No 3 Squadron with Lysanders was also operated from El Arish in an early stage of the conflict, but that claim seems a bit odd. The other REAF units did not participate in the conflict, at least until the autumn of 1948 when attrition proved to be higher than expected and risked the continuity of operations, this coinciding with the Egyptian initial belief that the campaign would be swift and soon turn to their favor.

The other Arab air arms involved were small and young and saw limited operational use during the conflict in comparison to the REAF. The Syrian Air Force (SAF) operated one squadron of armed NA T-6 Harvards in the light attack role from Almaza airport near Damascus, and they were little used in the northeastern part of Gallilee during the first two phases of the war. The Iraqi Air Force (IrAF) sent a small number of Avro Ansons, converted to light bombers, to aid the Iraqi expeditionary force. The Ansons were based at Mafraq and Amman in Jordan, and except for a small number of bombing sorties in the first and second round of the War of Independence, those Ansons were never seen in Israel's airspace. During the autumn of 1948 the Iraqis detached several Hawker Furies to Mafraq but did not take part in the conflict, although it was claimed that three Furies had also detached to El Arish, operating from there since October 1948. The Ansons and Furies' host nation was Transjordan, which at the time did not have an air arm, relying entirely for its air on the RAF that occupied Amman and Mafraq air bases according to a mutual treaty. The last nation to invade Israel during that crucial month of May 1948 was Lebanon, and its air arm was formed only by liaison and light transport aircraft.

While Ben Gurion announced the Independence Act at Tel-Aviv and the civilian population celebrated, less than five kilometers to the north, at the Sherut Avir's – Chel Ha'avir/IDFAF from now on – main base, some kind of other activity was taking place on that night. Six Hispano 20mm AA guns were deployed around Sde-Dov and the nearby Reading electricity power station. The aircraft were dispersed along the lone runway, and trenches were dug for the protection of the crews against air attacks. The activity ceased at around 0400 hrs in the early morning of Saturday, 15 May, and the tired men expected to get some urgently required sleeping hours, but it was exactly at 0520 hrs when Boris Senior, Sde-Dov air base commander, was woken up in his room at the Yarden Hotel in Tel-Aviv by the familiar sound of Rolls Royce Merlin engines and 20mm gunfire. Senior immediately drove his jeep to nearby Sde-Dov to find out that two waves of REAF Spitfires had attacked the airfield without any opposition since most of the AA crews were asleep. The damage inspection revealed a blazing ammunition warehouse, while the lonely hangar was slightly damaged, as was the runway, and several men were killed or injured. The uncamouflaged aircraft parked along the runway received the full attention of the Egyptian pilots. The largest targets were two DC-3s, an Air France machine and the PAAC ZS-AVK, and both were damaged. Pessah Tal was in the latter's cockpit when he heard strange noises:

A 101 Squadron pilot gets ready for a mission from Hertzlia airfield. The S199 was an unforgiving aircraft, and its massive propeller produced almost uncontrollable torque. (IAF)

"... In the evening of 14 May I was briefed at Tel-Aviv and I was supposed to take-off early in the morning from Sde-Dov to Tel-Nof, collect a crew and supplies, and fly from there to somewhere in the south. At about 0530 in the morning I was at Sde-Dov and I started one engine and it was running fine. Ithen started the second one and I heard noises like backfires. I first thought so, thus I shut it down but I still heard these noises!. Suddenly I heard the sound of the Egyptian Spitfires, there were four of them, and I finally understood the cause of the "backfires." I shut down the other engine too and fled from the Dakota, which was on the runway threshold ready to take-off and I ran into the sea with my clothers to take cover behind a nearby rock on the waterline. The Spitfires strafed the Dakota with cannons and machine gun fire. One cannon shell caused a 30 or 40 cm diameter hole in the vertical fin and many machine gun bullets struck the wings and fuselage. After the Spitfires left I consulted the IDFAF commander and an engineer. The latter said that the aircraft was unserviceable, but nevertheless we decided to fly it to Tel-Nof (ex-RAF Aquir) where aircraft shelters were available. I flew the aircraft alone, as it was, to Tel-Nof. There was no radio communication with Tel-Nof and I could not announce my arrival. To save time, as I feared that the Egyptian Spitfires might show up again and catch me in the air, I made a straight-in approach and landing and, as I stopped, a jeep drove beside the aircraft with Jacob Franck, the base commander, and Benjamin Peled, then a young mechanic (IDFAF C.O. in 1973-1977). They were armed with a machine gun and Peled walked with a Sten submachine gun ready to fire. They thought that the Egyptian army was landing to capture the field. It was funny, so we laughed a little and then I taxied the Dakota to a shelter."

An interesting story explains the fate of the other DC-3. The Air France representatives in Tel Aviv were officially informed that their aircraft was completely wrecked by the REAF attack. Air France collected the insurance compensation and the IDFAF gained a valuable aircraft soon to be repaired and pressed into service. A large number of light aircraft were also attacked by the strafing Spitfires. The Seabee was lightly damaged, and a Bonanza, Dragon Rapide and two Austers were seriously damaged, while a R.W.D.-

Chaim Laskov was the fourth Chel Ha'avir commander. This picture was taken during the sixties when he was in charge of Israel's Port Authority. (GPO)

13 (VQ-PAM) was completely wrecked. Senior ordered the dispersal of the remaining serviceable aircraft to emergency airstrips. The damaged Dragon Rapide was flown by Emmanuel Tsur, with mechanic Emmil Pohorita, to Haifa, where they discovered to their horror that a wing strut was broken. Three serviceable Austers were flown to emergency landing strips at the Sharon area, north of Tel Aviv, but it seems the dispersal had not been completed when the third wave of REAF Spitfires arrived at about 0800 hrs and the fourth some two hours later. It was the latter wave that suffered the first REAF loss in the conflict and offered the infant AA Corps its first victory. One Spitfire was damaged by 20mm AA fire when a shell hit the cooling system, and instead of flying east or south towards Arab held territory its pilot, Flt.Lt. Mahmoud Barakat, continued along the Mediterranean beach to the north. Boris Senior immediately took off in the serviceable Bonanza and followed the stricken Spitfire that made an emergency wheels-up landing at Hertzlia's beach, some 15 km north of Tel-Aviv. When Senior carefully landed the Bonanza near the damaged Spitfire, he found that the Egyptian pilot was all right and already under guard. He was interrogated about the REAF in general and about the El Arish air base specifically and confirmed the estimations of the REAF order of battle, although he claimed that a small percentage of the nominal force was serviceable. The downing of the Spitfire was a great moral achievement, but a war could not be won with moral and lessons had to be drawn. The mistakes were obvious: the Tel Aviv Sqn. aircraft were concentrated at Sde-Dov, which lacked any protective measures for aircraft except for very light AA fire; the aircraft were all civilian painted (uncamouflaged), and discipline and combat readiness was poor. Although an air raid was expected the air base commander was absent and only a small number of AA Corps men were on standby. Steps were taken to alter the situation and the aircraft were camouflaged, not uniformly as might be expected, but according to the paints available and the imagination of the painter. Nevertheless, most of the light aircraft were camouflaged in a green/brown scheme, although the tone applied to the Galilee Sqn. machines was much darker than the one adopted by the Tel Aviv and Negev Sqns. because of the darker northern terrain. On the other hand, as the larger transports were still required to fly abroad under fictitious covers, they were left in civilian colors and markings. The IDFAF insignia, a full blue star of David on a white circle, was also introduced after the nervous AA Corps crews fired at friendly aircraft. Senior painted a blue star of David on one of the aircraft and took off, but it could not be observed properly from the ground, so he filled it in with the same color and added a white circle as background to achieve a more satisfactory result that has not changed ever since. Because of the Arabs' total air superiority the light aircraft were instructed to fly only at night from 18 May onwards, but exceptions were not unusual due to urgent operational needs.

Sde-Dov was not the only target of the REAF on the first day of the conflict; other targets included the kibbutz settlements of Nirrim, Kfar Darom and Negba, and the Negev Sqn. base at Nir Am. In most cases the damage inflicted was light, but men were still killed and injured and equipment damaged. The REAF also attacked by mistake a British unit in the Negev and a soldier was killed, while eight others were injured. In the north, Syrian Harvards attacked kibbutz An Gev and supported the Syrian army forces attacking Massada in the Jordan Valley south of Lake Kinneret (Tiberias). The next day, 16 May, the IrAF joined the REAF and SAF. At 1130 hrs the Tel Aviv harbor was attacked by Egyptian Spitfires, while in the Negev, Nirrim and Nir Am were also attacked that day. The ex-RAF station of Tel-Nof was raided by Spitfires, and the Curtiss Commando NX67923, already grounded, was damaged and withdrawn from use for a considerable period. The Iraqi Ansons bombed Tiberius on the west bank of Lake Kinneret and Beit Yossef in the Jordan Valley, but caused little damage, while the SAF concentrated at the Massada/Sha'ar Hagolan zone to support Syrian forces that finally conquered the two points on the night of 18/19 May. A Negev Sqn. Auster was sent that day to bomb Egyptian forces north of Gaza, but the bombardier refused to fly as he feared REAF Spitfires. A Palmach platoon commander volunteered to be the bombardier and after they dropped their first 25-kg bomb they saw a pair of REAF Spitfires beneath. They dropped the second bomb and escaped into some convenient clouds, but it seems that the Spitfires followed the Auster and as it emerged from the clouds it was attacked. The Auster pilot managed to fly to the Negev Squadron at Nir Am where AA fire was opened on the two Spitfires and they broke away. Another Auster, flown by Menachem Bar of the Galilee Sqn., was also intercepted on 16 May by Syrian Harvards, but it too escaped without a scratch from the unequal air combat. 17 May saw the REAF attacking Tel Aviv (twice), Negba, the de-

Aviron was the birthplace of Israeli military aviation. In this picture taken on 25 October 1947 at Lod, can be seen most of its fleet, one Auster J/1, three Taylors, one Tiger Moth, two RWD.13s and one Be.550. (GPO)

serted ex-RAF air base at Kfar Sirkin about 15 km east of Tel Aviv (not used by the IDFAF as being too close to the Arab lines). Adam Shatkai, a Tel Aviv Sqn. pilot, on May 17, flew a replacement Auster to the Negev Sqn. at Nir Am and was welcomed by AA fire because REAF Spitfires previously attacked the place and the men were trigger happy. The next day, 18 May, saw the first area bombing in the Middle East since WWII when two Italian bombers had attacked Tel Aviv oil storage areas. Two Egyptian C-47s, flying undisturbed at a convenient altitude for best results, bombed Tel-Aviv's central bus station at dusk, causing the death of 42 innocent citizens. The Dakotas were neither challenged by fighters (there weren't any of them at the time) or AA fire, as the very small number of these guns protected military targets or supported the ground forces in various fronts as makeshift artillery. Tel Aviv was left unprotected to the mercies of the REAF. The following day saw the Arab air forces exploiting to the full their total air superiority, although one incident was to have a deep effect on the months to come.

It was a well known fact that while the formal British withdrawal date was 14 May, a small garrison still protected the Haifa harbor area until 1 August. The British forces were given air cover by the Spitifre FR.18s of Nos 32 and 208 Squadrons, which also were in the process to move to Cyprus from Ramat David air base, east of Haifa, while the Dakotas transported the groundcrews and sundry equipment from both squadrons. The RAF informed the REAF that they still occupied Ramat David, but the REAF's intelligence department did not bother to send this message to the operations department, and thus, in the early morning hours of 22 May, the usual formation of two Egyptian Spitfires appeared in the peaceful skies of the Izrael Valley and attacked the surprised Royal Air Force station. The attack was successful—two No 32 Sqn. Spitfires and a Dakota were destroyed on the ground while several others were damaged. The RAF immediately suspended the withdrawal and ordered reinforcements from Cyprus in order to maintain constant CAPs above the airfield. When the second and third waves of five REAF Spitfire 9s appeared, they were intercepted by the Superior Spitfire FR.18s of No 208 Squadron and the first Spitfire versus Spitfire air combat in the Middle East took place. The RAF claimed four downed by the CAP and another one by a Bren gunner, but before the actual interception these last two waves managed to attack, destroying a further Dakota as well as killing five RAF airmen and injuring several others. Two of the REAF Spitfires fell into Israeli-controlled territory, and the pilot of "624", Flying Officer Abed Al-Rahman Aiman, was taken prisoner by IDF forces. His Spitfire, which he managed to crash land, was salvaged for useful parts by the Israelis.

While the REAF, SAF and IrAF dominated Israel's skies in daylight, the IDFAF adopted the night and struck back with all it had, although it was not much. A Galilee Sqn. Auster flown by Peretz Grosser with Avraham Baltoy as bombardier dropped on the night of 15/16 May four bombs on a bridge on the Jordan river near Gesher where Jordanian forces concentrated for crossing the river. The bridge was missed, but a nearby IPC (Iraqi Petroleum Company) oil installation was set ablaze and burned for days, forcing the Jordanian armored forces to clear the area. The night flying order was ignored on emergencies and one of this was the evacuation of the women and children from kibbutz Beit Ha'arava in the Jordan Valley north of the Dead Sea. Following the demise of the four Gosh Etszion settlements, after the regular Jordanian army joined the irregular Arab forces attacking the besieged Gosh for several months, it was decided to withdraw from several points. One of those points was kibbutz Beit Ha'arava, and the plan was to evacuate the women and children by air and the men by boats to

Sdom on the Dead Sea. Pessah Tal and Cyril Katz left Tel-Nof in the damaged Dakota ZS-AVK for that purpose on 17 May, and Pessah Tal recalls:

". . . The hole in the vertical fin of the Dakota was patched with aluminum sheet and I was called to Tel-Aviv. Since the engineers feared that the machine gun hits had damaged the wing roots, they decided to remove the wings for inspection, at least a two-week work. At Tel-Aviv I was informed that women and children, evacuated from Beit Ha'arava, were concentrated at Calia, near the Dead Sea, and the Jordanian forces were on their way to that place. Nothing could stop them. The plan was to evacuate the men in a small ship. At Calia there was just a strip of land straightened by a bulldozer, and the largest aircraft that had landed there up to that day was a Dragon Rapide. I was asked if I could land there, load the women and children and fly them to Tel-Nof. I knew the place, I had landed there several times with light aircraft, and I said we could try. I was told that unless evacuated, the people there would be taken prisoner by the Jordanians. I flew with Cyril Katz in the Dakota as it was and we took minimum fuel with a reserve of less than 100 gallons to save weight. My plan was to make a pass on the field with the wheels up to watch what was going on down there and see if the strip was clear of obstacles.

I was on finals from the seaside when I observed a Jordanian column of armored cars about 15 or 20 km from Calia and moving. I decided to land without making the check fly-by so I told Cyril "gear down." In the end of the landing strip there was a small wooden shack. I came with power, full flaps and minimum speed, five or seven knots above stall. As I passed over the shack we closed everything, stalling into a three point landing. I stopped at the end of the strip and we started loading women and children without shutting down the engines. I did not count, I just let them in. The fuselage was empty, without seats or anything, to save weight and we took all the people present, including the engineer Novomaiski who was the director of the Dead Sea Works. I planned my take-off, which was against all the rules, but it was a Dakota and there is only one thing better than a Dakota and that is another Dakota. I opened full power, ignoring cylinder head pressure limits, I released brakes and after 20 meters I pushed the stick forward. The speed built on and at 80 mph, the stalling speed of a loaded Dakota, I gave quarter flaps and at 90 mph I was about 50 meters from the shack, but the wheels were still on the ground, so I lifted the nose a little and then pushed forward to gain speed. I pulled back a little and passed over the shack with the wings shaking on the edge of stall. After we cleared the shack, and above the Dead Sea, the speed built up and I was able to climb in circles. It took about 25 minutes to reach such a height that allowed me to cross the mountains between the Dead Sea and Tel-Nof, and at that time I saw that the Jordanians had reached the place. I flew back and passed over the burning Kfar Etszion. I had sentiments to that place since I risked my life there hundreds of times to evacuate wounded and fly in supplies. As we passed over Kfar Etszion I wept. It was sad. We landed at Tel-Nof and a bus waited for my passengers. I had evacuated 72 men, women and children in that flight. It was against the rules. It was luck. Anyway, they grounded the aircraft, removed the wings and discovered that the root was seriously damaged, and they had to rebuild half the airplane."

But daytime flying was dangerous, as experienced by a Galilee Sqn. pilot, Peretz Grosser, who on 19 May was intercepted in his Auster by a Syrian Harvard, which was armed with one forward firing 0.303-in machine gun and another such weapon firing afterwards and operated by the gunner. Grosser used the only tactic possible in such a situation: very tight turns up to ground level and then escape to the nearest air base. Grosser made it to the Galilee Sqn. base at Yavnial, but the Syrian Harvard followed him and the

38-year-old Israel Amir was the Chel Ha'avir's first commanding officer. (GPO)

At right is Aharon Remez, the second C.O. of the Chel Ha'avir. (IAF)

base was attacked the same day. Consequently, the squadron moved to the ex-RAF airfield at Ma'Hanaim, closer to the front lines, and observed from above by the Syrian forces camped on the Golan Heights, from where the unit mainly flew at night. During the daytime the four Austers and single Fairchild of the squadron were hidden among the trees, and although the SAF Harvards attacked the base several times, the camouflage was such that the aircraft were never damaged on those air raids. Furthermore, the SAF did not execute even a single night sortie during the whole conflict, and the Galilee Sqn. night activity was thus carried out undisturbed. On the south the Negev Sqn. suffered the same treatment from the REAF, and its base at Nir Am was twice attacked by Spitfires on 15 May and three times on 16 May, so the squadron moved to nearby Dorot and the activity carried on, despite the fact that hiding aircraft in the Negev desert was much more difficult than making the same in the green Galilee.

Both squadrons concentrated on local night bombing and the usual missions of evacuating wounded men to hospitals in Tel Aviv or supplying messages and urgent supplies to besieged settlements, although both those last missions were to suffer most from the Arab air superiority since night landings at those hazardous strips was obviously more dangerous. The Tel-Aviv Sqn. operating from Sde-Dov mainly flew in the central area of Israel, and it too concentrated on night bombing missions. On the night of 19/20 May three aircraft from this squadron, a Dragon Rapide, an Auster and a Bonanza, bombed in intervals the Syrian forces in Tsemah, just south of Lake Kinneret. On the night of 20/21 May, two Fairchilds bombed the Arab Village of Sho'afat in the vicinity of Jerusalem, a single Tel-Aviv Sqn. aircraft bombed Egyptian forces that concentrated at Dir Snid, south of kibbutz Yad Mordechai, which was under severe attack and was evacuated by its defenders on 23 May, while three other aircraft from the same unit, a Bonanza, a Dragon Rapide and a Fairchild, bombed in intervals Gaza airfield and the Egyptian army HQ at the Gaza railway station. On the night of 21/22 May, two Fairchilds bombed the Arab quarters of Sheich Ga'rah at Jerusalem. On the night of 22/23 May, aircraft from the Tel Aviv Squadron bombed the French Hill and Kfar Azaria in northern Jerusalem. On the next night the Negev Squadron bombed the Iraq El Suidan fortress, occupied by the Egyptian army, and also Majdal, both in the northern Negev, while the Galilee fliers bombed the Syrian army HQ at Kuneitra and Syrian forces at Kfar Hareb and Pik in the Golan Heights, well within Syria. On the night of 25/26 May the Tel-Aviv Sqn. bombed Jordanian forces in the Latron area on the Jerusalem corridor after an IDF attack failed, from which they suffered about 200 casualties. Three aircraft from the same squadron, a Fairchild, an Auster and a Bonanza, bombed Gaza once again. Also, during the night of 25/26 May a Dragon Rapide was sent to supply the besieged Jewish quarter in the old city of Jerusalem. The Jewish quarter defenders held at that stage an area of only a few hundred square meters and their situation was desperate. The Dragon Rapide was supposed to arrive at midnight and drop urgent supplies to the light of fires set by the defenders. Take-off was delayed because of mechanical malfunction and the Dragon Rapide arrived at 03:15 hrs (26 May) to drop five packages, in five dropping runs, from a 70 ft height. Two of the five packages reached the defenders, but the Jewish quarter finally surrended to the Jordanian army on 28 May. On the night of 28/29 the maiden Dakota bombing sortie in Chel Ha'avir service was executed. One of the Tel-Nof based Douglas bombed the Arab city of Rammallah, north of Jerusalem, with two tons of light bombs that were dropped from its door on the order of the pilot. Also during the night of 28/29 May, aircraft from the Tel-Aviv Sqn. bombed Nablus and Tul Karrem, while an attempt to bomb Janin failed when the pilot bombed instead She'an, which was already occupied by Israeli forces. Thus, while the Arab air forces controlled the daylight hours, the fledging IDFAF exploited to the full the night, and although night bombing is generally less accurate, especially with the primitive methods used, and the damage caused to Arab forces could be estimated to be small or very small, this effort raised the ground forces' morale. But while the three sectorial units fought their private night war, an activity which was about to change the situation took place at Tel-Nof air base, southeast of Tel-Aviv, codenamed "Oklahoma." In the afternoon of 16 May, two Curtiss C-46A Commandos left Catania and headed east towards uncertainty, Heyman Shamir navigated the lead aircraft, RX138, and the second, RX135, followed in close formation. Their destination was Tel-Nof, but the news from Israel was not encouraging and nobody knew if Tel-Nof was still occupied by the IDFAF as Arab propaganda announced the "great victories" of the invading armies. It was around 2200 hrs when the two large transports crossed the Tel-Aviv beach in close formation at 1,000 ft to be welcomed by AA fire from the nervous gunners at Sde-Dov. Shamir navigated the lead Commando to Tel-Nof, ignoring the Arab-held Lod International Airport, but radio contact with Tel Nof was not achieved and when the runway's lights were turned on they did not match the agreed code. The fuel was running low and Shamir accepted the risk: while RX135 circled above, RX138 landed, taxied to the runway's threshold and waited with engines running for any future developments. Heyman Shamir jumped off armed with a pistol and watched carefully the lone jeep that approached him, and when he finally recognized the two aboard, Aharon Remez and Yehuda Rabin, he knew he was at home. While the crews rested, the BT-13 was unloaded and the Commandos refueled for their return trip to Europe, Tel-Nof was considered too risky in the daytime as Commando NX67923 had been damaged there by REAF Spitfires the same day. It was still dark when the Commandos, freed from any burdening load for the first time since leaving the U.S., took off for a 14-hour flight to Czechoslovakia. The pair staged through Ajaccio where the night of 17/18 May was spent, arriving at Brno on 18 May. At this place the tired crews were instructed to fly to Zatec, a small airfield about 50 km from Prague, which was to be the center of Israeli operations from Czechoslovakia for the next three months. The vital supply airbridge was finally set up.

The Commandos were organized as a unit at Tel-Nof and gradually gathered from Panama and Brazil to augment those already operating on the "shuttle" to the central European country. A pair of leased Douglas C-54 Skymasters were also operated, and finally the Lockheed C-69 Constellation arrived from Panama in late June 1948 to join 106 Squadron at Tel-Nof, as the unit was known by then. The operations were planned and organized by LATA, and each flight was numbered in the Ballak sequence. Pessah Tal recalls that at the peak of the operations, each crew performed three round trips per week. The flight from Tel-Nof to "Zebra" lasted about 12 hours and was flown non-stop with auxiliary fuel tanks in the cabin. The flight route was west across the Mediterranean and then north to Czechoslovakia. Since oxygen bottles were not available, the flight was usually below 15,000 ft, but the Alps crossing needed to climb to this height or a bit higher for about half an hour, despite the lack of oxygen supply. The Commandos lacked IFR instruments and the crews avoided clouds by flying below their base, where the weather was worse, and the maps, according to Tal, were of unacceptable standards. The return trip evolved into a 14-hour affair through Ajaccio, since the loaded Commando could not cover the distance without a refueling stop. The Skymaster and Constellations flew non-stop both directions. Each Commando captain received $ 2,000 U.S. in cash for the crew expenses in Prague, for the refueling, for the bribery of local officials at Corsica, and for any emergency which should be encountered. The load was packed in crates on which "FRAGILE GLASS" was written in capital letters in order to deter curious officials at Ajaccio or anywhere

Canadian pilot Leonard Fitchett had a short career in the Chel Ha'avir, as he was shot down and killed in one of his first missions aboard one of the Beaufighters which had been purchased in the United Kingdom. (RCAF)

in Europe if the Panamian registered aircraft would have to make an emergency landing. After the four-hour flight to Corsica, the loaded Commando continued east across the Mediterranean to Tel-Nof, where it usually landed in the evening. It was unloaded, refueled, mechanically checked and flown back to Czechoslovakia by another crew, while the arriving crew was sent to Tel-Aviv for a rest. At first the aircraft were serviced at Zatec, where Yehuda Ben Hurin assumed command, and later, as the fear of REAF attacks reduced, at Tel-Nof, where Al Schwimmer was responsible for the maintenance of the large transports.

But, although the Commandos set the airbridge, it was a C-54 Skymaster that brought to Tel-Nof the first Avia S-199 on the night of 20/21 May after a non-stop flight which lasted eleven hours. Each S-199 was flown in one "Ballak" flight with its wing and propeller removed, and the equipment and ammunition required for operational use were also flown by the transports. The S-199s were accompanied by their intended pilots that had "graduated" from the conversion course in Czechoslovakia, as well as Czech mechanics that were to assist in the assembly at Tel-Nof. On the night of 21/22 May the second S-199 arrived in Commando RX 138, while on the next night the third one was unloaded at Tel-Nof from the Skymaster. Two more Commandos, RX 130 and RX136, arrived at Zatec to augment the operation and left for Tel-Nof on 23 May; the first carried bombs and equipment, while the second carried the fourth S-199, both flights being codenamed "Ballak 10" and "Ballak 11" respectively. The night of 23/24 May saw thick fog in the Tel-Nof area and fires were set in order to help the approaching Commandos. RX130 was the first and landed normally, but RX136 crashed into a hill not far from the air base. From its crew of five, four, including the two pilots Norman Monitz and Sheldon Icle, escaped unhurt, but the navigator, Moshe Rosenbaum, was killed. The Commando and the fighter it contained were both destroyed by the ensuing fire. Bad luck pursued RX130 as well; it encountered mechanical difficulties on the route, and an emergency landing was executed at Rhodes where the aircraft was interned, joining the also-interned Avro Anson, while its crew was arrested. After a couple of days of interrogation the crew was released, but the Commando had to wait in Greece for almost two years to be returned to the IDFAF. In spite of these losses the air bridge was by that time a regular operation with at least one flight per day. Up to the first truce on 11 June, which marked the end of the first round of fighting between Israel and the opposing Arab nations, about 107,000 Kg were transported from Czechoslovakia in 30 "Ballak" flights by the Commandos and the Skymaster. That load included eleven Avia S-199 fighters, associated equipment, and small arms.

The original plan was to wait until five Avia S-199s would be ready at Tel-Nof and then to perform a surprise attack on the REAF forward air base of El Arish. The number five was in fact the number of fighter pilots available: Modi Allon, Lou Lenart, Eddie Cohen, Milton Rubenfeld and Ezer Weizman. For that purpose the Egyptian pilot, Flt Lt Barakat, was interrogated by Aharon Remez and Ezer Weizman in order to obtain the maximum available information about El Arish. Aerial photography was out of the question because of lack of reconnaissance aircraft. Moreover, in order to preserve surprise the Avias were not flight tested—the maiden flight from Israeli soil of those hybrid German/Czech fighters was also to be their baptism of fire!. Each assembled fighter was carefully camouflaged by nets to hide it from the REAF eyes, while the five nervous pilots waited for their first mission. On 26 May, Holda, a kibbutz near Tel-Nof where the IDF HQ for the Jerusalem corridor was located, was attacked by REAF aircraft and Ezer Weizman jumped into one of the S-199s. He started the engine and obviously wanted to intercept the attacking Egyptians, but he was refused permission and thus a vehicle was used to block his path. Heavy pressure was exerted by the General Staff to attack with only three S-199s on Friday, 28 May, but it seems that the IDFAF staff managed to ignore that pressure. Later on that day Remez reported to Ben Gurion:

"... The Knives (S-199 call-sign) were not ready at 0100 (28 May). The bombs were not ready and we do not have an armorer. There was something missing in the bombs. The Czech technicians could not load the guns."

And on the morning of 29 May: "... The Messerschmitts did not attack. The bombs, the fuel and the brakes were not ready. They will be ready at 1100."

Needless to say, the S-199s did not take-off at 1100, and it may well be that somebody was doing his best to wait for the fifth Avia since four had been assembled by Saturday, 29 May.

Nevertheless, and in spite of all those real or imaginary malfunctions, it was on that Saturday, 29 May, that the Avia S-199s finally took off, although there were only four of them and the target was not El Arish. Since that morning the besieged strong point of Nitzanim, some 45 km south of Tel-Aviv, reported continuously about the advance to the north of a huge Egyptian army column. A small bridge 32 km south of Tel-Aviv was blown and the head of the column was concentrated near it. Between that bridge and Tel-Aviv was the Giva'ty Brigade, tired after two weeks of fighting and lightly armed in comparison to the Egyptian forces. The commander of the Giva'ti Brigade asked for air support, and since the Tel-Aviv Sqn. light aircraft could not attack during daytime the only means to carry out that demand was the four S-199s at Tel-Nof. Aharon Remez, IDFAF Head of Operations and the man behind the El Arish attack plan, opposed the infantry forces' demand, but as Nitzanim kept reporting, it was obvious that a drastic effort was required to stop the Egyptians. Finally, Yiga'el Yadin, the IDF Head of Operations, accepted the Giva'ty Brigade demand and Shimeon Avidan, the Brigade commander, rushed to Tel-Nof with the order to attack the Egyptian column (which was 20 km south of Tel-Nof) and brief the pilots. It is of interest to note that a similar air support demand by Shlomo Shamir, the commander of the hard-pressed Seventh Brigade facing the Jordanian army at Latrum on the Jerusalem corridor, was rejected the same day. The pilots quickly organized, and they decided to operate in two pairs. Lou Lenart and Modi Allon would take-off first, followed shortly by Ezer Weizman and Eddie Cohen. The foursome bombed and strafed the Egyptians at and near Ishdod (see the opening of this book for the combat report), but Eddie Cohen, hit by flak, crashed and was killed near Hatzor air base (ex RAF Station Qastina), probably trying to crash land on the airfield, although it is possible that Cohen mistook Hatzor for Tel-Nof (ex-RAF Aqir) since Hatzor is only about 12 km southwest of Tel-Nof, then on Arab-held territory.

His body was found a year later after the Egyptians withdrew and the area was searched. Modi Allon wrote-off his S-199 during landing, although he was unhurt. It could not be stated that the strike was a success since one pilot and two aircraft were lost and considering the fact that only four aircraft and five pilots were available at that stage it is fair to say that attrition was high. Furthermore, the S-199 showed at its debut two major shortcomings—the two cowling mounted machine guns, which fired through the propeller arc using a synchronization mechanism, were totally unreliable, and the narrow landing gear track plus the large propeller made every landing a hairy story. In several cases pilots that fired the guns damaged the propellers, and as it revolved to the right the hard torque produced many landing and take-off accidents, adding to the lack of rudder trim. The Avias did not have at that stage an oxygen system and were restricted to a height of up to 14,000 ft. Moreover, the groundcrews not used to this equipment found that when the S-199s landed at Tel-Nof that only half of the fuel tanks had been filled up!. On the credit side was the fact that the Egyptian column stopped its advance to the north, and Tel-Aviv was saved. In the coming days the Egyptian forces were attacked at Ishdod by the Giva'ty Brigade and several other units, and although all these attacks failed, the Egyptian's army 2nd Brigade lost the initiative and Ishdod was finally liberated in October 1948. Thus, the first combat appearance of the S-199 in Israeli skies was not in vain. The Egyptians did not suffer severe damage from the eight 70-Kg bombs dropped by the quartet of obsolete fighters. The damage sustained by the strafing runs which followed the bombing was estimated to be small, but the Egyptians were surprised; its high staff had not considered the possibility of true fighters on the other side as realistic, although their forces were well equipped with light AA weapons, as witnessed by the four Avia pilots at dusk on May 29, 1948.

In order to maintain pressure on the Egyptian column, the Tel-Aviv Sqn. was instructed to attack the Ishdod area during the night of 29/30 May. Furthermore, an empty C-46 Commando on its way

Rare picture of a Beaufighter operating in Israel, as these fighter bombers were rarely operational during their short IDFAF career.

from Tel-Aviv to Zatec bombed , the Egyptian forces at the same place before dawn on May 30. With the Egyptian army held on defensive lines at Ishdod and the El Arish air base attack abandoned due to the lack of surprise and the fact that only two S-199s remained, the Avias were also used on other fronts as well. An Iraqi light armored column attacked Kfar Yonah, east of Netanya, on 30 May, and threatened this town just 30 km to the north of Tel-Aviv. The capture of this place would mean that Israel would be cut into two main parts, the Galilee in the north and the Tel-Aviv area in the center, while the Negev and the Jerusalem areas were already isolated by the Egyptian and Jordanian armies, respectively. The two remaining Avias were the only aircraft capable of safe daytime operations and both were sent to attack the Iraqi column on the early morning hours of 30 May. It was 0530 hrs, less than twelve hours after the first attack, when the Jumo-engined fighters took off from Tel-Nof. Ezer Weizman led Milton Rubenfeld with the same load as on the previous day—two 70-Kg bombs each plus full internal ammunition for the two fuselage-mounted MG 131 13mm machine guns and the two underwing mounted MG 151 30mm cannons. They bombed the railway station at Tul Kar'rem and strafed the column on the road. Rubenfeld's S-199 was damaged by AA fire or from the explosion of his own bombs, and he was destined to be the first IDFAF pilot to bail out of his aircraft. Luckily he managed to fly to Kfar Vitkin, north of Netanya, and jumped over the beach, being injured in the process. Weizman's luck was no better since he suffered a bird strike and his windshield was broken. Weizman landed safely at Tel-Nof and a few holes were also found in his S-199 wings. The next morning Weizman broke his left arm in an accident while on his way from Tel-Aviv to Tel-Nof on Modi Allon's motorcycle and the status of the small fighter outfit deteriorated even more to only two pilots, Allon and Lennart, and one damaged Avia, but the same day a fifth machine was delivered with its assembly process completed. By that time Mordechai "Modi" Allon was appointed as the first C.O. of the unit, soon to be known as 101 Squadron (The First Fighter Squadron), and the light gray S-199s began to be painted with red and white striped rudders, used up to this day by 101 Sqn. in its F-16C/D Fighting Falcons, and the squadron insignia, a white skull with black wings on a red circle, was painted on the left side of the nose. On 31 May the urgent plea for air support from the Seventh Brigade at Latrum was finally answered when Modi Allon escorted, in the only remaining S-199, a Tel-Aviv Sqn. Dragon Rapide to bomb the Jordanian forces in the area. Meanwhile, the Arab air forces were still very active at that stage. On 29 May the REAF attacked kibbutzs Gvar'am and Nir'am in the Negev, and one of its Spitfires was seen leaving the area trailing smoke. IrAF Ansons bombed Israeli forces facing the Iraqis east of Netanya. On the following day the Ansons returned, bombing Affula and Merchavia in Izrael Valley. On 1 June the REAF attacked Tel-Nof air base and four other targets in the Tel-Aviv area, including the IDF staff camp at Ramat-Gan, and on 2 June two REAF Spitfires attacked Holda while others attacked Negba, in the northern Negev, and Rishon Letsion, Gan Yavne and Sarafend barracks, all south of Tel-Aviv. The light aircraft of the Tel-Aviv Sqn. retaliated with the usual night bombing. On the night of 30/31 May Ramle and Lod were bombed, and on the following night, Israeli aircraft attacked for the first time an Arab capital. The members of the Arab League met on that evening at Amman, Jordan, to discuss the developments of the war, and it was decided to bomb the city as a show of strength. Three aircraft were sent in short intervals from Sde-Dov. A Bonanza, a Fairchild and a Dragon Rapide. Ben Gurion noted in his diary on 2 June:

". . . The RAF at Amman informed its HQ at Haifa that the RAF station was bombed last night at 0300-0315 hrs. Four bombs were dropped, three fell within the camp and one outside the fence. Three incendiary bombs fell in the airfield. No losses. A second attack was on the city. Ten killed and 30 injured. One vehicle in the airfield was damaged and one hangar destroyed. Two Ansons were damaged."

On 3 June the IrAF bombed Sdom from high altitude, while the REAF bombed the IDF General Staff at Ramat Gan at 1130 hrs and attempted to bomb Tel-Aviv at 1900 hrs. It was attempted because this time a lonely Avia S-199 took off from Tel-Nof and to the eyes of thousands of astonished citizens – it was not widely known that Israel possessed fighters – shot down the two unfortunate REAF C-47s, one after the other. One crashed south of Tel-Aviv in Israeli held territory, and the second still further south in Egyptian controlled territory. All the crewmembers of the two aircraft were killed. The S-199 pilot was Modi Allon, and it is only significant that it was an Israeli pilot that was credited with the first air-to-air victory in IDFAF story. It was the last time that REAF C-47s attempted to bomb Tel-Aviv; nevertheless, the Spitfires continued operating, and although the damage sustained during their attacks was small, the light aircraft of the sectorial squadrons were still restricted to nighttime operations only.

Encouraged by its first aerial victories, the IDFAF retaliated intensively during the night of 3/4 June with the Tel-Aviv Sqn. performing most of the bombing sorties that concentrated that night on two areas, the Egyptians in Ishdod and the Iraqis east of Netanya. Three aircraft were sent to bomb Ishdod. The first was a Dragon Rapide that took off from Sde-Dov at 2040, dropped 300 Kg of high explosive bombs and 18 incendiary bombs and returned to base at 2145. Next was a Bonanza that left Sde-Dov at 2055 and bombed Ashod with 50 Kg of HE bombs and 12 incendiary bombs, landing back at 2145. A Fairchild Argus was the third and last aircraft that left Sde-Dov that evening to Ashod. It took off at 2100 but could not identify its target and dropped its bombload, 100 Kg of HE bombs and six incendiary bombs, into the Mediterranean Sea before landing at 2215. The Iraqis were attacked later that night and four aircraft participated in the action. Two Dragon Rapides, a Bonanza and an Argus were sent towards Nablus and Tol Karrem, and all aircraft returned, landing at Sde-Dov between 0625 and 0650 hrs. Also, the Tel-Aviv Sqn. performed two supply missions to Sdom and Jerusalem, respectively.

Following the Douglas C-47 losses, the Egyptians tried to shell Tel-Aviv from the sea on the afternoon of 4 June. Israel did not have a navy that could fight the Egyptian navy out in the open sea and many feared that the shelling would be followed by an amphibious landing. The IDFAF was called on for help, and three Tel-

Aviv Sqn. aircraft were tasked to attack the ships: a Dragon Rapide, a Bonanza and an Argus. Clearly this was a desperate mission, as they had to operate in daylight, without fighter cover, and against well defended and difficult targets in the form of warships. Boris Senior was the first to attack at 1530 with his Bonanza armed with two hand dropped 50-Kg bombs and two externally carried bombs, mounted under the fuselage. As expected, Senior missed. The second to attack, five minutes later, was the Fairchild Argus, followed by Senior's second sortie in the Bonanza at 1610. Again the Argus returned five minutes later as well as a Dragon Rapide, piloted by Benjamin Inbar, making a single attack at 1620. A third sortie by the Bonanza followed at 1640, and finally the Argus attacked at 1655. In that last sortie the Argus was downed by AA fire and its crew, Aharon David Shprintsak and Matityaho Sokenio, were killed. The seven attack sorties by the three aircraft, were not in vain however, since the Egyptian ships ceased the shelling and retreated, and one of them was damaged.

A Syrian Air Force Harvard was shot down by Israeli AA fire at Ayelet Ha'shachar on 5 June. These converted trainers also attacked that day Rosh Pina and Malkia, while the IrAF bombed IDF forces at the Netanya area and Affola. On the night of 5/6 June the Tel-Aviv Sqn. operated as usual, but this time bad luck intervened and at 0545, while four aircraft were loaded with bombs, an armorer, carrying a 25-Kg bomb, struck the revolving propeller of a Fairchild Argus. The bomb exploded and the Argus, already loaded with two 25-Kg bombs and 24 incendiary bombs and about to receive two more of the former, was completely wrecked. Two armorers and the Argus bombardier were killed, while the pilot was severely injured. A nearby Dragon Rapide, loaded with twelve 25-Kg bombs, was badly damaged, and two more armorers were also wounded. Another Fairchild Argus was struck off the inventory when it was damaged in a take-off accident while serving with the Galilee Sqn., though the crew survived the roll over. This accident took place on the night of 10/11 June.

On 6 June, the SAF Harvards attacked Mishmar Ha'yarden, Rosh Pina and Machanaim, and the IrAF bombed the Affula area. The REAF's new trend to support the army was shown on 7 June when kibbutz Nitzanim was attacked by Spitfires in support of an armor and infantry push that ended with the Israeli stronghold surrounding that day. On 8 June a REAF Spit was downed south of Tel-Aviv, and its pilot, Flying Officer Jamal Orfan Safe El-Nasser, was captured by Israeli forces—the victor was Gideon Lichtman flying an Avia S-199, and thus getting the type's third kill. The next day three Spitfires attacked Rehovot, also south of Tel-Aviv. On 10 June the SAF attacked three Israeli strongpoints, and on 11 June, the last fighting day of the first round, the REAF attacked kibbutz Ro-hama in the Negev seven times.

Some days previously, on 24 May, RAF Station Ramat-David was finally handed over to the IDFAF, and Joshua Gilloch was appointed as the new air base commander. The first flying unit to arrive was the Galilee Sqn., following the shelling of its base at Machanaim by Syrian artillery. The squadron move to Ramat-David was after the usual night activity. Each pilot flew to the new base following his last sortie on that night. The first was Moshe Peled, the squadron C.O.; he had bombed Syrian forces south of Kinneret Lake and then flew to Ramat David. The night was cloudy, so Peled flew above the clouds until he found a hole in the cloud layer and descended to find himself above Janin. He landed first and was followed by the other Austers of the unit. Silloch gave Peled the rank of Rav Seren (Major) since he was a squadron commander, but the Galilee Sqn. could not adjust themselves to the military discipline imposed by Gillock. He organized Ramat David as a real military air base with separate dining messes for officers, NCOs and enlisted men, and with ranks and salutes to the officers. The squadron men, accustomed to the standards of the Golani Brigade, could not handle the situation, and two versions about their departure from Ramat David exist. Gillock claims that he expelled the squadron out of the base, while the Galilee Sqn. men insist that they "escaped" from Ramat David. Anyway, the squadron moved back to its original base at Yavniel and many stories still survive about that departure, including a claim that the squadron bombed the Ramat David headquarters with rotten tomatoes!

Wreckage of Fitchett's Beaufighter after being shot down by Egyptian flak, he and his gunner were killed.(IAF)

In early June, the two Dakotas, based at Tel-Nof up to that time and little used, were flown north to Ramat David to find a new unit, 103 Squadron. Danni Rozin, an ex-SAAF pilot, was appointed as the C.O. of the new squadron, and its first significant operation was the bombing of Damascus. This raid was planned for the night of 9/10 June, but mechanical malfunctions delayed the operation to the next night. It was a full moon night and the Dakota loitered above the Syrian capital for half an hour while dropping two tons of bombs. The SAF Harvards did not intercept while the AA fire was scarce. The actual damage was probably small since the bombs were thrown out of the Dakota's door by "bombardiers", but it boosted the Israeli morale and degraded the Syrian's at a critical stage of the conflict—the evening of the first truce. As the first truce came into effect, one fact was clear—Israel survived. The Negev was still detached from the center of the new state by the Egyptian forces in the Falluga area, while Jerusalem, the intended capital, was still besieged by the Jordanian army, but the initial advance of the Arabs was halted, and their victories fell short of their expectations. The campaign was not swift and short as they had expected, and although they obtained successes, the battles were fierce and their losses high. The IDF, on the other hand, was encouraged by the first round's results. Almost all the planned operations of that period failed, but the invading Arab armies were stopped and the confidence within the IDF grew to such an extent that the second round was opened by an Israeli initiative.

For the IDFAF the end of the first round meant reorganization; the service had entered the War of Independence on 15 May with about 900 men, of which 50 were aircrew, 350 ground crew and 500 in administration, logistics and staff. Between 60 and 80 men were recruited per week in the period following 15 May, and the number of enlisted men grew steadily to more than 2,000 by the end of July. Experienced foreign volunteers occupied almost every job. A group of 13 Israelis that were given flying training in the U.S. also returned in early June to augment the number of pilots of the sectorial squadrons. While the number of enlisted men grew steadily, so did the number of aircraft. On 15 May some 30 lightplanes were operated by the sectorial squadrons. During the first round the first IDFAF fighters and transports arrived. One leased Douglas Skymaster and five Curtiss Commandos joined the lone unserviceable Commando at Tel-Nof to create a regular service to Czechoslovakia. Eleven Avia S-199s were airlifted to Israel by these transports until 11 June. The lonely leased PAAC Dakota was joined by the Air France machine, and both were sent to Ramat David to create the nucleus of the future transport squadron. An Australian Douglas DC-5 (VH-ARD) that was resident at Haifa was bought on 5 June for 12,000 Israeli Pounds, but saw very little service. The light aircraft were augmented by three Dragon Rapides, a Miles Gemini and a Miles Aerovan. All were procured by Emmanuel Tsur in England and flown to Israel without any problems under fictitious identity. The losses of the first round were three Avia S-199s, a Curtiss Commando, two Fairchild Argus, a Noorduyn Norseman, and one R.W.D. 13. A Commando was interned in Rhodes and another was unserviceable at Tel-Nof. Two Dragon Rapides, one Beech Bonanza, one Republic Seabee and one Avia S-199 were damaged and grounded for repairs, while another two unassembled S-199s were damaged by a REAF air raid on Tel-Nof on 1 June. The losses in aircrews were four pilots killed and another one severely injured, while one bombardier and one navigator were also killed. The C-64 Norseman mentioned was lost on 20 June in Urbe airfield, Rome, while Leonard Cohen gave refresher training to George "Buzz" Beurling, when the aircraft crashed after exploding, possibly due to sabotage by Arab agents. Both pilots were killed, and the loss of Beurling, a Canadian who served with the RAF in WWII and was credited with 32 aerial victories, mostly over Malta, was a major blow.

By the end of the first round the IDFAF had gained valuable experience and its units occupied five air bases. But, as might be expected, not all was well. The serviceability rates were low due to a lack of spares, and skilled ground crews and the fact that aircraft suffered damage in almost every operational sortie. The relations within the IDFAF were tense, and in the afternoon of 17 May a delegation of three pilots met David Ben Gurion to inform him that the IDFAF structure was wrong and that it lacked internal cooperation. Boris Senior, the Sde-Dov commander and one of those pilots (the two others were Gideon Elrom, the Tel-Aviv Sqn. commander and Pessah Tal that had returned that morning from Beit Ha'araval) suggested to invite Cecil Margu, a Jewish South African who served as a Wing Commander with the SAAF during WWII, as a consultant, and surprisingly enough Ben Gurion agreed. On 22 May Margu was officially invited as an aviation consultant to the Israeli government, although he only arrived in Israel exactly two months later on 22 July.

Although possessing a relatively large staff, the IDFAF could not handle the airbridge to Czechoslovakia, and this operation was conducted by LATA under the leadership of Meir Mardor who, as already mentioned, was an excellent organizer without any aviation background. Even on the operational side difficulties were encountered, and the relations between the IDF General Staff and that of the IDFAF were tense over this issue; there was still a long run ahead for the Chel Ha'avir to become the most efficient air force in that part of the world.

Shlomo Shamir, the third commander of the Israeli Air Force photographed during the late sixties. (GPO)

Chapter 4:
First Truce and Ten Days

The first truce, sponsored by the United Nations, started on 11 June and was to last for 28 days. These four weeks were vital for the IDF and the Chel Ha'avir in particular. The lessons of the first round were learned the hard way, and thus those four weeks were very welcomed for reorganization and training. The reorganization was required because of the constant influx of aircraft, equipment and volunteers, and the lessons from the operations during the first round of the conflict. But the reorganization involved at that period only the field units, while the staff structure was not affected during the first truce and the ten days of fighting that followed before the second truce was announced on 19 July.

As already briefly mentioned previously, the two available Dakotas, the Air France machine and ZS-AVK, were moved to Ramat David for 103 Squadron under the command of Danni Rozin, a volunteer from South Africa who served as a bomber and transport pilot with the SAAF. This pair was joined by two others, ZS-AVO and ZS-DSC, from South Africa, while a fifth Dakota, NC-74136, was donated to the IDFAF by the Jewish millionaire from Holland, Bernard Van Leer, whose wife worked at the time at besieged Jerusalem as a nurse, and while his son, Wim, volunteered to serve with the IDFAF as a pilot. 103 Squadron received only four Dakotas and used them for internal transport flights and bombing missions. The fifth Dakota was used to fly senior officials to Europe from Tel-Nof. The unit that was engaged with pure military service to Europe was also based at Tel-Nof and was numbered 106 Squadron. Its workhorse was the Curtiss C-46 Commando, but it also operated a pair of leased Douglas C-54 Skymasters, which were later bought, and a single Lockheed C-69 Constellation, and it was under the direct control of LATA, who organized the airbridge to Czechoslovakia. The Commando fleet included four serviceable machines (RX133, 135, 137 and 138), while RX132 (ex NX67923) was still unserviceable due to damage sustained by AA fire over Yechi'am before 15 May, and a REAF Spitfire attack on Tel-Nof on 16 May. Meanwhile, RX134 and RX136 were lost. RX130 was still at Rhodes, and the remaining two, RX131 and RX139, arrived from Panama during the first truce to augment 106 Squadron, also augmented by the Constellation, RX121, which arrived at Zatec from Panama on 25 June; its pilot on the ferry flight was Sam Lewis.

Up to the first truce the aircraft operating "Ballak" flights brought to Israel 107 tons of vital equipment, including eleven Avia S-199 fighters, small arms and ammunition. The operation was feasible due to the refueling stop at Ajaccio, Corsica, but as the first round ended, political pressure, especially from the United States and Great Britain, forced France to impose limitations on Israeli activities in its territory. Ajaccio airport was closed to LATA's operations from 11 June onwards. On 15 June Yugoslavia agreed to allow Israeli aircraft a refueling stop within its borders, and thus the vital airbridge continued. Moreover, the refueling in Yugoslavia meant a reduction in the distance from 4,000 kilometers through Ajaccio to 2,900 Kilometers via Yugoslavia, shortening the flying time from fourteen to ten hours. This refueling stop was required by the Commandos on their way back from Czechoslovakia. The Skymasters and Constellations flew non-stop both directions. During the truce the unit lost two aircraft. Commando RX133 landed at Tel-Nof on the night of 3/4 July as "Ballak 55", it was unloaded and refueled for its return trip.

A few T-6 Texans or Harvards were available to the IDFAF late in the War of Independence, and were used as dive-bombers and close-air support. (IAF)

Wreckage of the Royal Egyptian Air Force Spitfire serial EP601, shot down by Modi Allon on 18 July 1948 near Be'er Sheva. The picture was taken four months after the dogfight. (IAF)

The refueling procedure at Tel-Nof was very primitive at the time due to lack of equipment. The fuel was pumped from barrels by manually operated pumps. As the left engine was started, the left wing caught fire, probably due to a careless refueling procedure, and before the firemen arrived, also equipped with unsuitable equipment, the Commando was engulfed in flames. The crew, who were all Swedish, escaped unhurt, but the aircraft was lost. The second accident involved the Constellation, and although it was not totally lost, the aircraft returned to full flying conditions only after the War of Independence ended. The accident took place on its seventh "Ballak" flight on 14 July. The Constellation was on its way from Tel-Nof to Zatec with about twenty passengers, including Heyman Shamir, who was on his way to Czechoslovakia to escort the B-17s to Israel, and empty ammunition boxes. On approach to Zatec the hydraulic system failed and only the left landing gear was lowered. The crew consisted of Martin Ribcoff, an American volunteer as the captain, and Misha Kener, an Israeli who served as a pilot in the Red Army during the Great Patriotic War as the copilot. Kener tried to retract the left landing gear in order to execute a belly landing, but it refused to retract and all the attempts either to lower the nose and right landing gears or retract the left one failed. Sam Lewis, who watched the drama from the ground, took off in a Bultee BT-13 that was resident at Zatec and tried to push the left landing gear up with the BT-13's wing. That attempt failed, too, and finally Ribcoff and Kenner had no option but to land. The crash landing was performed excellently, and although the aircraft was severely damaged, it left Zatet on 18 September 1949 for repairs at Amsterdam, and the crew and passengers escaped unhurt. The Constellation accident took place only four days after the U.S. Customs service and FBI interned the two other Constellations—the primary cause of this internment was because of the B-17 affair.

The Boeing B-17 Flying Fortres was on the shopping list from late 1947. Surplus B-17s were used in the U.S. for a variety of roles and there were no difficulties buying them. The difficulty was to delivery them to Israel. Al Schwimmer organized the operation and bought four B-17s and four Douglas A-20s. Two B-17Gs, NL-1098N and NL-5024N, were bought from Charles T. Winters for U.S.,30,000, one B-17G was bought from Universal Aviation for U.S.$15,000, and the rest were obtained from the War Assets Administration for about U.S.$40,000. Winters used his pair of Flying Fortresses as freighters between Miami and Puerto Rico and Universal used the B-17 for aerial photography in California. The next stage was the refurbishment of the aircraft and recruiting crews to ferry them to Israel. At a very early stage it became evident that refurbishing the surplus aircraft would be lengthy, and it was decided to smuggle the three airworthy B-17s first. Leo Gardner organized the flight. When Gardner returned to the U.S. from Europe in March, he staged through the Azores and arranged there that for a cash payment of 1,000 Sterling Pounds, his aircraft would be refueled and serviced. The B-17s were now to test this agreement. Another arrangement was with Ajaccio; although that airport was closed for LATA's aircraft, the French were asked for one more favor. The plan was that the B-17s would announce at the Azores that their destination was Ajaccio. The B-17s would continue directly to Zatec and the French were asked to confirm the Azores that the B-17s landed there in order to avoid a search for three "lost" aircraft. A search that would be mentioned in all newspapers and turn the U.S. authorities' attention to Schwimmer's activity and the aircraft still in the United States. The three B-17s were gathered at Miami and took off for Puerto Rico on 12 June. Schwimmer and Gardner saw the first two depart, but the third suffered a flat tire and left Miami a few hours later. At Puerto Rico they filed flight plans to Brazil because an export license was not required for such a flight. Instead of flying south, the Fortresses headed east to the Azores, where Gardner's arrangements worked well, but they encountered a thick fog. Nevertheless, the crews decided to continue immediately and on 14 June the three B-17s landed at Zatec. The arrangement with the French did not work, and it was published worldwide that three B-17s had disappeared over the Atlantic. The press soon revealed the truth, and in order to minimize the damage, Sam Lewis was ordered to fly the Lockheed Constellation from Panama to Israel. The Constellation was left at Panama for repairs

Typical accident of an Avia S199 from 101 Squadron. During its short operational life, rarely were there more than two or four Avias in operational conditions, but the truth is that it was the only fighter available during several months. (GPO)

and it waited there for the pair of Constellations that were refurbished in the U.S., but by now it was clear that there was no chance to smuggle those large transports and it was feared that the strong U.S. influence in Panama would force the Panamian authorities to intern the aircraft. Lewis landed RX121 at Zatec on 25 June and the final blow came on 9 July when RX123 and RX124 were interned by the American authorities, although the B-17 smuggling story had an epilogue.

On 11 July Leo Gardner flew the fourth B-17 from Tulsa, Oklahoma, to Westchester, where he was replaced by Irwin Shidler, the ex-president of Service Airways, who filed a flight plan to the west coast but flew instead north to St John, New Foundland. Bad weather forced him to land at Darmouth, and the crew was arrested by the Canadian authorities that found a load of small arms and spares in the aircraft. The U.S. authorities demanded the turning in of the aircraft and crew and the Canadians released the aircraft on 17 July, but it was given fuel in the amount required for the flight to Boston. The Flying Fortress never arrived there, and early the next morning, 18 July, it landed at the Azores. Once there the tired crew made its only mistake, and instead of continuing immediately as did the three B-17s a month earlier, they rested for a couple of hours and American pressure forced the Portuguese authorities to arrest them and intern the aircraft. The crew was turned in to the U.S. and the B-17 never left the Azores. In March 1949 Irwin Shindler was sentenced to prison on parole, a light punishment compared to that of Charles Winters, who was sentenced in February 1949 to 18 months in prison. Also, in February 1950, both Schwimmer and Gardner were sentenced to a U.S.$10,000 fine each for their part in the B-17 affair.

The three Flying Fortresses were parked at Zatec and their crews joined the airbridge operation, flying Commandos, Skymasters and Constellations. The young IDFAF air staff was not able at the time to handle the airbridge operation, for which purpose LATA was founded, and it seems that these four-engined bombers were beyond its capabilities. Moreover, the three B-17s were not combat ready: the machine gun turrets were removed and faired over, the oxygen systems were deleted, the bomb racks and bombing sight were missing, and the three aircraft could hardly be claimed as bombers at that stage. For three long weeks the Flying Fortresses were ignored by the IDFAF air staff, and Ray Kurtz, LATA's operations officer at Zatec, appointed mechanics to guard the grounded planes and proposed himself to the air staff as the commander of the B-17s. Kurtz, an American volunteer who served as a B-17 squadron commander with the USAAF during WWII, was ignored by the air staff, which was hard-pressed by the rapid expansion of the IDFAF. Finally, towards the end of the first truce, the air staff ordered Kurtz to assume command of the B-17s and ready them for combat operations within five days. It was 7 June when Ray Kurtz received his orders and it seems that the panic was due to the realization by the air staff that the first truce was about to end with an all-out offensive by the IDF, for which massive aerial support was required by the ground forces. The first truce was to end on 9 July, and it was clear that the B-17s would not be ready on time, so the air staff asked the IDF General Staff to postpone the offensive until their arrival; this was refused, and the IDFAF had to do without them during the first seven days of the Ten Days fighting.

Ray Kurtz gathered the B-17 crews, and together with the ground crews, they worked around the clock. It was an enormous job, but on the sixth day, 13 July, the three B-17s were "ready." As might be expected there were not any B-17 original subsystems in Czechoslovakia, and the men had to improvise with what they had—WWII surplus German equipment as well as Czech equipment based on outdated German designs. Since electrically powered gun turrets were not available, each aircraft was equipped with three machine guns for manual use. One was located in the tail, where originally the tail turret was, and the other two were installed in the waist positions. The machine guns were manually operated by the gunners, and the only type available were the Czech BZ37, also

Three B-17G Flying Fortress were smuggled out of the U.S., and they formed number 69 Squadron, nicknamed "The Hammers". (IAF)

Late in the war a few P-51D Mustangs were available, with only two operational, serving with 101 Squadron, which after the war was completly reequiped with the type until the mid-fifties when this supreme fighter was replaced by the Dassault Mystère IVA. (IAF)

101 Squadron pilots beside one of their Spitfires. Note the censored squadron badge. (IAF)

known as Beza; less reliable and with a lower firing rate than the original 12.70mm Brownings. The bomb racks that were installed were also of Czech/German design and enabled the B-17s to carry four 250-Kg and eight 70-Kg bombs that were supposed to be delivered in daytime from high altitude, and for this task two more subsystems were missing; the original Norden bombsight and the oxygen supply system. Only one of each (non-originals) were obtained in Czechoslovakia, originally installed in the same aircraft.

But now let's turn the attention to 101 Squadron, by then the only IDFAF fighter outfit. This unit, which commenced on 29 May, was left with only one serviceable aircraft by the next day, and the lone Avia S-199 was kept on interception readiness at Tel-Nof until the end of the first round, achieving the first of six victories on 3 June. Between the appearance of the Israeli fighters and the first truce, the REAF discovered that their base was at Tel-Nof and Spitfires attacked it on 1 June, damaging a pair of unassembled S-199s and their hangar. As a consequence it was decided to remove 101 Squadron from Tel-Nof and relocate it in one of the landing strips that were erected in the Sharon area, north of Tel-Aviv. Modi Allon and Ezer Weizman chose the one in Hertzlia, just east of Kfar Shmarriaho and roughly 15 km north of Tel-Aviv, where a 2,000-meter unpaved runway was constructed by bulldozers in a south-north heading among the orange groves. A nearby water tower, that can still be observed actually, was used as a control tower and the aircraft were parked and serviced under camouflage nets among the orange trees. Weather the REAf was aware of the activity at Hertzlia or not, it did not attack the new base even once. For the same reasons that forced 101 Squadron to Hertzlia, the Central Maintenance Unit (CMU) responsible for the assembly of the Avia S-199s at Tel-Nof, was relocated near kibbutz Ma'abarot in the Sharon, some 50 km to the north of Tel-Aviv. Eleven fighters arrived by air from Czechoslovakia during the first round, five during the first truce and the rest during the Ten Days. By the end of the first truce the IDFAF had 14 Avia S-199 fighters of which four/five were claimed to be combat ready in the hands of experienced and highly motivated pilots. To augment Modi Allon, Ezer Weizman, Lou Lenart and Milton Rubenfeld, a number of newly arrived foreign volunteers were sent for conversion in Czechoslovakia. Most of them were Americans, but four came from South Africa and one from the United Kingdom, and these were among the group that arrived in Israel towards the end of the first truce. The Americans included Rudi Augarten, Chris Magee, Robert Lester Vickan, Sid Anton and Leo Naumis. The South Africans were Sydney "Syd" Cohen, Arnold Roch, Lionel Morris Bloch and Boris Senior, while the Brit was Maurice "Murray" Mann. When the new volunteers arrived 101 Squadron was already at Hertzlia, and they found an organized unit by the then operating standards. Although some of the foreign volunteers were more experienced than Modi Allon, he retained his position as the C.O. while Maurice "Murray" Mann and Sydney "Syd" Cohen were appointed A and B Flight commanders respectively, and Rudi Aurgarten the squadron's operations officer. Harry Axelrod, an American volunteer, was the technical officer, and Kalman Turin, an Israeli who served with the RAF during WWII, was the administration officer, and both helped to run the daily routine in the squadron that consisted of some 250 men in all. Boris Senior, who was the Sde-Dov air base commander before the conversion to the S-199 at Czechoslovakia, was appointed as an operations officer with the air staff upon his return to Israel, but kept on flying occasionally with 101, which had by the Ten Days Period an aircraft/pilots ratio of one-to-one, although the number of serviceable S-199s was quite low—usually below four available fighters at any given time.

101 Squadron pilots in May 1949: from the background, right to left, Rudi Augarten, Steve Baker, John McElroy, Murray Mann, Sid Cohen, Ezer Weizman, Danny Wilson, Cesar Dangott, Sandy Jackson, Arnold Roch, Sid Antein, Bill Pomeranz, Jack Doyle, Cy Feldman, Lee Sinclair and Wayne Peake, a very talented —most having served previously in the USAAF, RAF, RCAF and SAAF— a cosmopolitan group! (IAF)

Two views of Spitfire Mk.9 coded 664 of the REAF after being captured by Israeli ground forces at El Arish airfield in Sinai late in 1948.(IAF)

The Avias were disliked by the pilots since all of them were used to British and American fighters. Ezer Weizman, 101 Sqn. third commander, claimed that they chose Hertzlia partly because they considered that the S-199 would behave better from an unpaved runway. Sid Cohen, the second C.O., thought that the unpaved runway at Hertzlia caused some of the take-off and landing accidents there. Both, like Augarten, Neumis, Rubenfeld and the rest, were extremely unhappy to fly the S-199 in combat and all hoped to fly better fighters, a wish that materialized when a lone Spitfire was brought on a semi-trailer to Hertzlia one day. That Spitfire was the first of a pair that were reconditioned from the remains of British and Egyptian aircraft. During the last days of the British mandate, attempts were made to bribe British officials in order to lay a hand on RAF aircraft, much in the same way as the Haggana obtained two Cronwell and one Sherman tanks at Haifa. However, these attempts failed, and all serviceable aircraft were evacuated from Palestine by the RAF, which deliberately damaged all unserviceable machines, engines and equipment that was left behind at its base. After the British evacuated and the Haggana forces moved in, a search was started by the Sherut Avir, and later Chel Ha'avir, for worthy equipment. Those search groups found a large number of derelict aircraft and because of the desperate situation—it was before 15 May—it was decided to focus on the refurbishment of fighters. It was also attempted to recondition a single De Havilland Mosquito, although that work was later abandoned, and on 22 June Israel Amir, IDFAF C.O., asked for a special budget of 8,000 Israeli Pounds for the refurbishment of Vickers Wellingtons that were at Tel-Nof, and that the technical department assumed that two of them could be brought to an airworthy condition. By this time the B-17s were already at Zatec and the Wellingtons were never refurbished.

The Spitfire remains that were also found at Tel-Nof were transported to the Sherut Avir's maintenance center at Sharona and work commenced under the command of Freddy Ish-Shalom. The aircraft that were found were of different marks, and the reconditioned one was a hybrid, modified for the recce mission by the Israelis. Merlin engines were found at both Tel-Nof and Ein Shemer, while Zachi Shaul, a mechanic who worked on BOAC Lancastrian engines at Lod Airport, overhauled one engine. On 16 May Barakat's Spitfire was brought from Hertzlia's beach to Sharona, and a week later the remains of REAF Spitfires shot down by the RAF were also gathered at Sharona. Finally, in late June or early July, the first Spitfire coded D-130 (in June 1948 the IDFAF implemented a serial system in which each aircraft received a letter that indicated its type: A, light; B high capacity light aircraft; D combat and S transport, followed by a number, which were allocated in batches, thus the S-199s began at D-101, the Spitfires at D-130, the Mosquitos at D-160, the Beaufighters at D-170 and the Mustangs at D-190), thus the recce configured Spit—the camera had been installed in the empty oxygen tank bay with a hole cut in the bottom of the fuselage to expose the lens, while a small control panel was installed in the right side of the cockpit, with a green light illuminating when the camera was activated—was assigned to B Flight of 101 Squadron, led by Sydney "Syd" Cohen, and only experienced Spit pilots were assigned for the recce missions, the first batch included Cohen himself, the Squadron CO Modi Allon, and A Flight CO Maurice Mann, and the pilots Leo Nomis and Arnold Ruch. On 5 August Syd Cohen was the pilot making the first flight with the reconstructed aircraft, while the same day Maurice Mann flew the first two operational sorties, the first to Faluja and the second to Be'er Sheva. On 8 August Mann made a test flight to see if the camera worked and then flew the first two operational photo sorties. A total of 16 recce sorties were flown during August, and in two of these missions D-130 was lightly damaged by AA fire. Meanwhile, a second Spitfire constructed from scrap was taking shape—it was D-131.

A Curtiss C-46 Commando flying over the Haifa Bay with the oil refinery in the background.(IAF)

The changes that had swept in the IDFAF during late June and early July did not overlook the light aircraft units. The Tel-Aviv, Negev and Galilee Squadrons were numbered 1, 2 and 3 Squadrons respectively, although in this narrative the acronyms that reflect their sector of operations will be used throughout. The small number of light aircraft that were added to the inventory during that period included: a pair of Grumman Widgeon amphibians donated by the Duchman Bernard Van Leer, together with the Dakota NC-74136; four of the light-planes bought by Emmanuel Tsur in England (a pair of De Havilland Dragon Rapides, a Miles Aerovan and a Miles Gemini) that arrived on 10 June; and the slow but constant flow of Norsemans arriving from Europe. As already indicated, Cohen and Beurling were killed in one of these at Urbe airfield, Rome, on 20 May. On the second circuit the aircraft exploded, and it was speculated that a time bomb, smuggled into the aircraft by Arab agents, caused the explosion. The C-64 Norsemans that did arrive were assigned to the Tel-Aviv Sqn. and were used to supply remote points such as Sdon on the southwestern corner of the Dead Sea. However, on 27 June the French finally surrended to American pressure and interned seven Norsemans. Simultaneously, the Dutch and Belgian authorities interned the remaining Norsemans still in Europe and it seemed like the end for Fredkins' aircraft, but counter pressure was applied, forcing the release of some of the aircraft, which were flown to Zatec to wait there for the ferry flight to Israel. Abraham "Abe" Nathan, an Indian Jew who flew with the RAF in WWII and later became famous for his Voice of Peace ship-mounted radio station, and Richard "Grisha" Brown, a Russian Jew who flew fighters in WWII with the Red Air Force, collected a

Norseman in Toussus le Noble to ferry it to Israel. They were interred at Nice by the French, but released after three weeks. Instead of flying to Israel they proceed to Zatec, but lost their way and landed in West Germany, managing to arrive to the Czech airfield the next day. Far less lucky were Robert Fine and William Mulfin, who ferried a pair of Norsemans to Israel in early July. They reached the Israeli coastline, but it was completely blacked out, so they flew along it until lights in an airfield were turned on and they both landed there. It was Gaza! The two men were the first IDFAF aircrew to be taken prisoners of war.

In spite of the truce, the light-plane units were busy with supply, liaison and observation missions, plus training flights to improve the skills of veteran pilots and to bring to operational status inexperienced pilots that were kept on ground crew jobs as future reserves. Some of the best light-plane pilots were allocated to other units to gather experience on heavy types since it was well understood that one day the foreign volunteers would leave and the IDFAF would have to count on the young Israeli pilots. Pinchas Ben Porat and Misha Kener joined the heavy freighters at Tel-Nof, while Benjamin Inbar, Adam Shatkai and Jacob Ratoshniac were sent to Ramat-David to fly as copilots in Dakotas. The first truce did not mean that violations did not occur and clashes took place from time to time when Israeli forces attacked Arab garrisons within Israeli held territory and vice versa. The Arabs attacked besieged Israeli points that were in Arab-held territory. On the night of 12/13 June the Galilee Sqn. bombed two hostile Arab villages in the Galilee, Tor'an and Lobiya, that bothered Israeli forces. The Galilee Sqn. diary also reveals some of the above-mentioned activities: on 16 June a night observation flight was executed to familiarize five young pilots that had joined the three veterans, Moshe Peled (the CO), Peretz Grosser and Yesha'ayahu Bodilovski, in low flying, night flying, STOL operations, etc. Finally, on 30 June a formation flight was practiced. On 25 June Tsvi Zibel of the Tel-Aviv Sqn. was sent to supply the Besieged Ben Shemen strongpoint, in the Jerusalem corridor, and for that flight he was awarded the highest Israeli decoration, Ot Ha'gvora (The Gallantry Decoration). Only twelve such decorations were awarded during the War of Independence, and Zible was the only airman to receive it. The award read:

"...31145 (personal military number) Airman Zibel Tsvi showed gallantry when he risked his life in the battlefield at Ben Shemen on 25 June 1948 and this is what he did:

31145 Airman Zibel Tsvi left Sde-Dov to deliver supplies and ammunition to besieged Ben Shemen. During that time enemy fire forced the defenders to take cover. He landed at the narrow strip and without any help brought his load to a secured place. All that was done under constant enemy fire, which became heavier and thus he could not return to his aircraft. According to his proposal, the defenders loaded sand bags on a tractor, and he drove it alone to the open field and parked it between the aircraft and the enemy. He succeeded in starting up the engine of his aircraft, which suffered several hits from the intense small arms fire, and returned safely to his base."

Zibel was later killed when his Auster was shot down by REAF Spitfires in the Negev in December 1948.

On 26 June two Tel-Aviv Sqn. Austers were damaged, the first due to a technical malfunction at Hertzlia and the second by Arab fire at Ben Shemen, but both pilots survived. A fortnight later another Auster from the same unit was lost and its pilot killed. On 7 July Amnon Berman was sent in for a photographic sortie around Revivim, in the Negev, and later on the same day to bring men from the Negev to Tel-Aviv. Above Lod he was downed by ground fire and the Auster crashed in Arab held territory. Previously, Berman headed a group of thirteen Israelis that studied at universities in the U.S. and volunteered for a flight training course at Bakersfield, California. The members of the group received about 200 hours each and were sent to Israel during June. Berman himself arrived via Panama in a ship that delivered ammunition to Israel.

By the end of the first truce the General Staff thought that the IDFAF had matured and thus, since the Ten Days, were to be opened by an all-out offensive on all fronts, and naturally expected to receive its full cooperation and support. On the central front Operation Danni was planned against Jordanian and local forces in the Ramle/Lod area, east of Tel-Aviv, to ease the pressure from Jerusalem and remove the Jordanian threat on Tel-Aviv. In the north, Operation Brosh (a kind of tall tree) was planned to drive away from Israel the invading Syrian forces, and in the south, although the IDF remained static in the first few days of the Ten Days, it later launched Operation Ma'wet La'polesh (Death to the Invader), meant to open the road to the besieged Negev. All three operations required massive aerial support and the General Staff considered that the available Chel Ha'avir order of battle could satisfy those requirements, although its air staff recommended to wait for the B-17s. However, as can be expected in a war-like situation, not everything always develops as planned.

Pilots of 101 Squadron pose beside one of their Avia S199s early in August 1948: on cowling are Murray Mann, Ezer Weizman and Aharon Finkel. Bottom row - Sid Cohen, Chris Magee, Bill Pomerantz, Giddy Lichtman, Sandy Jacobs, Leo Frankel, Sid Anton and Leo Nomis. (IAF)

Chapter 5: The Second Truce

On 9 July the fighting began with an IDF offensive in the north, against the Syrian forces occupying Mishmar Ha'yarden, and in the Jerusalem corridor against the Jordanian forces in the Ramle/ Lod area. In the south the IDF was to hold out against the Egyptians until reinforcements would make an offensive feasible. For the first time in the fighting the IDFAF was ordered by the General Staff to execute a sort of pre-emptive attack on El Arish air base at dusk on 8 July, in order to put pressure on the REAF and achieve local air superiority over the front lines for the first day of fighting. Other IDFAF missions were air raids on Iraq, El Manshiah, and Falluja in the northern Negev, where Egyptian forces were concentrated, on the night of 8/9 July. However, all these operations either failed or were canceled. The four Avia S-199s tasked with attacking El Arish were not ready in time and the foursome were only ready in the early morning hours of 9 July. Of the four "Knives" that were planned to attack El Arish, one crashed on take-off from Hertzlia, though luckily enough the pilot, Alexander "Sandy" Jacobs, escaped unhurt, and the remaining three S-199s proceeded to El Arish, but they failed to locate it and instead attacked Egyptian army targets around Gaza and its harbor. However, only two Avias returned home; Robert "Bob" Lester Vickman, an American volunteer who had served with the USAAF in the Pacific as a recce pilot, was listed as missing in action aboard aircraft D-110, possibly shot down by flak.

The bombing of the Syrian HQ was canceled because there were Dakotas available, and both the bombings of Iraq, El Manshiah, and Falluja failed: instead of bombing the former, kibbutz Ro'hama was bombed without any casualties to the IDF forces occupying the place, and the bombs addressed for Falluja found their way to the Mediterranean Sea because the pilot failed to locate the target.

The REAF retaliated on 9 July with Spitfire attacks on Tel-Aviv and IDF strongpoints in northern Negev. Finally, on the dawn of the same day, 9 July, the Galilee Squadron commenced a series of attacks against hostile Arab villages in the central Galilee in conjunction with Operation Dekel (a tree found mainly in deserts) that was to start on 10 July with the aim of freeing the central Galilee from hostile forces and neutralizing the threat on the Israeli western Galilee. The Galilee Sqn. dropped during that dawn and dusk 22 bombs, for a total weight of 332 Kg on five bombing sorties. All the sorties were performed with the Austers, and on each, four-

Wreckage of the RAF's Tempest F.6 shot down by Bill Shroedder on 7 January 1949 near kibbutz Nirim.

pound incendiary bombs were also dropped, while on some of the villages leaflets were dropped calling for the local population to surrender. The Syrian army HQ at Kuneitra was finally bombed in the early morning hours of 10 July by a single Dakota from Ramat-David, and on his way back the pilot had the wish to support the hard-pressed ground forces in the Mishmar Ha'yarden area, but visibility was poor and the remaining bombs were dropped at Nazareth in the central Galilee.

Later on that day, 10 July, Syrian Harvards supported their forces in the Mishmar Ha'yarden area and attacked the Operation Brosh HQ at Rosh Pina, vehicles on the Rosh Pina-Machanim road and IDF strongpoints. The Harvard attacks were effective and caused damage and a large number of casualties. The Northern Front HQ requested from the General Staff air cover, and a pair of Avia S-199s were scrambled from Hertzlia to the north. The 101 Sqn. fighters encountered two SAF Harvards, and an air combat followed in which the Israeli leader Maurice "Murray" Mann, shot down one of the Syrians while his wingman, Lionel Morris Bloch, chased the second Harvard well into Syrian territory and disappeared. Bloch, a South African fighter pilot who flew with 4 Sqn., SAAF, in Italy during WWII, became the second 101 Sqn. pilot listed MIA. Following Vickman the previous day, Bloch was flying D-107 and his remains were returned to Israel in 1949. In two days of fighting 101 Squadron had lost two pilots and two aircraft, plus a third badly damaged for only one kill achieved by the unit's "A" Flight C.O., having totally failed to negate REAF air superiority.

Flying Officer Timothy John McElhaw, RAF, which bailed out of his Spitfire FR.18 after being shot down by Slick Goodling of 101 Squadron on 7 January 1949. (IAF)

During the same day the Egyptians attacked IDF targets in the Negev continuously, including Ro'hama, which was struck by seven waves of Spitfires, Negba, Nir-Am and Be'er Tovia. But the trend of the last days of fighting before the first truce was followed: only the Spitfires participated in the fighting, and they concentrated in tactical support missions in the Negev area, completely ignoring IDFAF bases and executing a small number of "strategic" attacks on targets in central Israel. Those later attacks caused little damage because all were carried out by the Spitfires, while the C-47s disappeared from the daytime scene and the number of REAF night operations registered during the War of Independence did not reach a dozen missions. While Operation Brosh in the north did not achieve its aim to drive the Syrians away from Mishmar Ha'yarden, Operation Danni in the central front was gathering momentum, although it was given only negligible air support. A planned sortie to bomb Anaba on the night of 9/10 July by a single Tel-Aviv Sqn. Dragon Rapide was canceled because the aircraft was not ready in time, but nevertheless that village was captured the next day by the Iftah Brigade that forced its way to besieged Ben Shemen. Also, on 10 July Lod International Airport, closed for air traffic since April 1948, was captured by an armored regiment equipped with a pair of British Cromwell tanks and a small number of pre-WWII vintage French Hotchkiss light tanks. When David Ben Gurion arrived at Lod airport on 14 July for a visit he wrote in his diary:

> "....The airport—who knows if the Israeli government could have constructed such an airport in ten years—is in our hands. It is almost undamaged. I saw the two heavy tanks—Cromwells—that we inherited from the British. They overtook the airport."

11 July saw the first Arab air raid on Jerusalem when a single REAF aircraft dropped several 100-Kg bombs, and three children were killed. REAF activity that day was intense, and the El Arish-based Spitfires attacked many IDF strongpoints in the Negev, including the Brigade HQ and Negev Sqn. base. Following the previous day's failures, Israeli air activity was quite restricted, although a S-199 formation was sent to the north for a combat air patrol in the Mishmar Ha'yarden area. That day no Syrian Harvards were intercepted, but Sydney "Syd" Cohen, on his first operational sortie, fired a short burst to test his cowling guns and damaged the propeller due to a defective synchronization mechanism, although he made it back to Hertzlia.

The Egyptian army regained the initiative, and on 12 July launched a massive attack on kibbutz Negba with the usual support of REAF Spitfires. The attack failed, but other targets in the Negev were strafed and bombed. Iraqi Ansons appeared from Jordan to bomb Rosh Ha'ain, just north of the area covered by Operation Danni, which had been captured from the Iraqi forces by the Alexandroni Brigade. Meanwhile, the IDFAF failures caused frustration among the General Staff, and on 13 July, in a formal document, it was claimed that, although the Air Force had 2,500 men and some 30 operational aircraft, it failed to execute the General Staff orders. It is true that it was suffering at the time from a makeshift organization, and a low serviceability rate due to lack of spares, equipment and skilled ground crews, and poor cooperation level with the ground forces. Nevertheless, all the same applied to the ground forces, too, and although they faced failures a lot, they managed, through high motivation and determination, to overcome the difficulties while the Chel Ha'Avir did not.

That evening, 13 July, saw yet another IDFAF failure when a single Dakota was sent from Ramat David to bomb the Syrian forces at Mishmar Ha'yarden. At that time the Dakotas were not equipped with external bomb racks and the primitive bombing method was to drop the bombs out of the open door. The bombs were passed, from hand to hand, until the last "bombardier" at the door pulled out the safety pin and threw the bomb in the darkness. The pilot monitored the bombing by the red/green sequence usually used for paradropping and the faster the bombardier worked, the number of required passes to drop all the load and time above the target decreased as a consequence. That evening, the seventh bomb struck the Dakota's tail, but the pilot managed to fly back to Ramat-David where the almost uncontrollable aircraft performed a crash landing without injury to the crew, but the aircraft was a write-off. The next day on 14 July, a pair of Arab aircraft supported an attack on Seg'gera in central Galilee, but the Golani Brigade pushed back the attackers. On 15 July Iraqi Ansons bombed Affula and Meggido in Izrael Valley, and it was on the same day, more than 2,500 km away, when an operation that was to completely change the IDFAF fortunes was commenced at Zatec, Czechoslovakia. The plan was to bomb Cairo (Amman and Damascus had already been attacked), and the mission was called Operation Nabut (a kind of stick used for fighting) involving the three B-17 Flying Fortresses.

Following the Ray Kurtz effort to recondition the four-engined bombers in five days, Heyman Shamir was sent by the air staff to Zatec on 14 July to learn first hand the situation there. He found out that the crews were afraid of being intercepted by the RAF fighters based in the Suez Canal Zone and thus it was decided that instead of bombing Cairo with all three bombers, only the lead B-17, equipped with an oxygen system and a bombing sight, would attack Cairo, while the other Flying Fortresses were assigned alter-

Many of the lightplanes which served with the Sectorial Squadrons kept their civilian colors through the war. (IAF)

native targets: Gaza and El Arish air base. In the last two targets no air opposition was expected, since the REAF lacked any night interception capability, and both were much closer to the Israeli lines than Cairo. The lead aircraft was captained by Ray Kurtz and Heyman Shamir accompanied him, the rest of the crew being: William Katz, copilot; John Gross, bombardier; Harvey Nachman, navigator; Don Albretch, flight engineer; Abe Axelrod and Marvin Price, gunners—there was also a radio operator aboard. The captains of the other two bombers were Norman Monitz and Al Reisin, both veteran Ballak pilots. The three B-17s left Zatec on the morning of 15 July for their maiden operation in IDFAF service. The original plan was to bomb Cairo at dusk since there was not any information on whether the Egyptian capital was blacked out at night or not, but it turned out that the Egyptians were so self-confident that they did not even consider the possibility of an air raid on Cairo. The lonely B-17 dropped eight 250-Kg and eight 70-Kg bombs on several targets within Cairo, including the Egyptian army officer's school and the king's palace. Except for a few searchlights that tried to locate the Flying Fortress, no other opposition was encountered.

According to Egyptian sources the bombing caused 85 casualties, of which 30 were killed. The other two B-17s were less successful and probably both bombed Rafah, in northern Sinai, instead of Gaza and El Arish air base. As the three bombers landed at Tel-Nof a debate within the IDFAF was started; some argued that since the REAF might attack and destroy the Flying Fortresses on the ground they should be returned to Zatec and operate from there as "shuttle bombers", while others claimed that the B-17s should be operated within Israel for maximum efficiency. A compromise was finally adopted to base the three bombers in three separate air bases and thus reduce the risk of losing them all. The B-17s were dis-

North American Harvards during the 1948 War. Note the identification markings used during that period, as the Syrians also used the same type of aircraft.(IAF)

persed to Ramat-David, Aben Yehuda and Tel-Nof, but it was soon realized that for efficient control, operation and servicing, the three should operate from a single base as a unit. On the morning of 16 July, Dan Tolkovski of the air staff flew with the three bombers to bomb El Arish. After the S-199 failed attempts to strike at this important EAF air base in the Sinai during the past weeks, El Arish was finally bombed with seven and half tons of bombs. The B-17s' tactic was to fly low to the target area in order to avoid detection and then climb for the bombing run. During the low level flight to and from the target, the gunners used their weapons to strafe Egyptian positions. Since fighters were scarcely available, no escort was provided at that stage and the Flying Fortresses had to rely on the defensive power of their Beza machine guns. They were never intercepted by Arab fighters during the War of Independence, despite the fact that on the majority of the missions they operated during daytime and at medium to low levels. Adding to the efficiency of these heavy bombers was the fact that because of the short distances involved—usually the radius of action was less than 200-km—a considerable load could be carried. But the B-17s had their share of troubles, too: they relied on Czech and Russian-made bombs, of which the stocks were low and consumption high due to the above-mentioned reason. As a consequence, during one of the bombing missions on the Syrian forces at Mishmar Ha'yarden, the B-17s had to use the popular hand bombing method with mortar shells!. Yet another difficulty was that the bombers had been designed for high altitude operations, and in order to exert the maximum from them at low altitude sorties the crews had to improvise solutions. Nevertheless, the B-17s were extremely reliable by IDFAF standards of the period and most of the sorties were executed with all three bombers flying in formation. At noon on 16 July, the Flying Fortresses bombed the Egyptian forces at Ashdod and Majdal, and finally at dusk two of them bombed the Syrians at Mishmar Ha'yarden. By the end of that day the B-17s were all at Ramat-David under the command of Ray Kurtz. The unit was organized as 69 Squadron, nicknamed "Ha'patishim" (The Hammers), and less than a week later Kurtz returned to LATA and Bill Katz succeeded him as C.O. The B-17s changed the IDFAF fortune almost overnight, but the Arab air forces were still active, and on 16 July, Iraqi aircraft attacked an armored IDF column that overtook Nazareth in central Galilee and supported the Jordanian counterattacks on the Bir Main and El Borge strongpoints in Operation Danni, although both counterattacks failed. A strange incident took place on the night of 16/17 July when the B-17s were ordered to bomb Damascus and the Syrian forces at Mishmar Ha'yarden. Both missions were not carried out because Joshoua Gilloch, the Ramat-David air base commander who previously tackled the Galilee Sqn., insisted on receiving from the air staff a clear definition of the troop's positions in the front line (what is called today FLOT) for the Mishmar Ha'yarden bombing and defined targets for the Damascus raid. The air staff did not answer Gilloch as he expected and thus he canceled the mission! David Ben Gurion wrote about this incident in his diary on 17 July:

"....The night bombing of the Syrian camp and Damascus was not executed, according to Yiga'el (Yadin, Head of the Operations Department, IDF General Staff), because of the interference of Gilloch, and I gave Yiga'el the authority to arrest Gilloch (if necessary)."

It seems that Gilloch was a real problem since as early as 13 June Ben Gurion wrote in his diary:

"....(Emmanuel) Tsur visited Ramat-David and although he had a letter from Chel Ha'avir (probably means the air staff) Gilloch did not let him stay at the air base (according to Tsur)."

The B-17s bombed Mishmar Ha'yarden in the morning of 17 July while Damascus was bombed by two Dakotas separately a few hours earlier. The Dakota that bombed at 0300 hrs, 17 July, stayed above the Syrian capital for 45 minutes and dropped two tons of bombs without any interference. The Syrians lacked any means for night interception and moreover, the city was not blacked out. Also on the same day the Fortresses turned their attention to Nablus and Tl Karren, where Iraqi forces were resident, dropping there five tons of bombs. While the new bombers received all the limelight, the light-planes of the sectorial squadrons continued with their unglamorous but invaluable tasks of light night bombing, liaison, observation and casualty evacuation. On 17 July a Tel-Aviv Sqn. led by Miles Aerovan was sent to Sdom, on the shores of the Dead Sea, to bring in supplies and evacuate several wounded soldiers and bring back for rest a few more. On its way back it suffered an engine failure and force-landed south of Bat Yam, a town south of Tel-Aviv. The pilot, Andrew Boyd Spencer, an American volunteer and four of the passengers were murdered by a tribe of Bedouines, while two of the passengers managed to escape. The Aerovan was set ablaze by the nomadic Arabs.

Since noon on 16 July it became evident that the U.N. would announce a second truce, to come into effect at the Jerusalem area in the morning of the next day and in the rest of the country in the morning of 19 July (both at 0545)—it was obvious that 18 July was

Edward "Eddie" *Cohen, a former fighter pilot with the SAAF in WWII, flew with one of the sectorial squadrons before being posted to 101 Squadron in May 1948. He was shot down and killed by Egyptian flak on 20 May as the first for the new fighter squadron. (IAF)*

the last full day of fighting for that round. Furthermore, since that second truce was not a restricted period like the first truce, it was clear that both sides would try to achieve the best they could on that last day of fighting. In the north following the failure of Operation Brosh the fighting against the Syrians in the Mishmar Ha'yarden area turned static, while in the central Galilee the success of Operation Dekel removed the threat on western Galilee, although further attacks failed. A similar situation was in the central front. The success of Operation Danni's first stage did not lend to the capture of Latrum, on the Jerusalem corridor, due to the fierce resistance of Jordanian forces in the area. Nevertheless, on all these fronts the IDF held out just fine—the problem was in the south, and the limelights on 18 July turned to the Negev. Operation Mawet La'polesh failed to secure an open corridor to the besieged Negev, and the Egyptian forces retaliated with counter attacks supported by the REAF Spitfires. At 0600 in the morning of 18 July the B-17s were sent to bomb Egyptian forces at Majdal and on the same morning a B-17 bombed Damascus and its airport, El Maza, where the SAF Harvards were based. Later on that same day the B-17s returned for yet another attack at Majdal, and a single Dakota bombed Kalkilia and Tul Karrem on the central front. It was the first daytime Dakota air Raid, and it only shows the growing confidence within the air force following the appearance of the Flying Fortresses and their effective operations.

In the afternoon of 18 July three Avia S-199s were sent from Hertzlia to attack Egyptian forces engaged with the IDF at Bir Aslog in the Negev, south of Be'er Sheva. Modi Allon, 101 Squadron C.O., led two American volunteers, Rudy Augarten—on his first operational sortie with the IDFAF—and Sydney Antin. They made three strafing runs at the Egyptian forces, and on their way back at a height of 6,000 ft they observed a pair of REAF Spitfires positioning for a bounce on the Avias. In the air combat that followed, which was for all intents a true dogfight, Augarten stayed in despite having exhausted his ammunition during the previous strafing passes. Modi Allon shot down one of the Spitfires, whose pilot, Wing Commander Mohammed El Janzuri, was killed. It was Allon's third kill in less than two months. When the formation landed back at Hertzlia at 1800 hrs, Allon and Antin slightly damaged their Avias, but both escaped without injury.

A Jeep is loaded trhough the side cargo door of a C-46 during Operation Avak in August 1948. Note that the Panamanian civil registry has been covered by black paint.(IAF)

The activity of the Galilee Squadron during the Ten Days period was quite important, mainly supporting Operation Dekel, where Arab air activity was negligible, and thus freed the heavier aircraft to support IDF forces facing the Syrians, Jordanians and Egyptians. This unit exploited to the full the favorable environment denied to the other two sectorial squadrons, which operated under constant REAF threat, and in a period of sixteen days in early July it recorded 91 bombing sorties and dropped a total of 6,025 Kg of bombs., all with less than five Austers and ten pilots. No wonder that Moshe Carmel, the commander of the northern front, thanked that squadron on 21 July:

"Number 3 Squadron Commander Northern Front Headquarters

General Staff/Ops Department AD/9/332
Chel Ha'avir's staff 21 July 1948

As operations ceased I want to express my appreciation to the efficient work of the squadron under your command and the good support we received from that squadron in the numerous operations within the Northern Front since the end of the truce. I admire the full cooperation between the ground forces and the air force that made feasible the success of our work. I am full of gratitude to the pilots under your command that efficiently executed the operations with so few means.

M. Carmel
Commander of the Northern Front"

The paragraph about the "full cooperation" between the ground forces and the air force must be considered cautiously as a sectorial success.

The real turning point came with the arrival of the Flying Fortresses on the evening of 15 July. In the three days that followed the B-17s operated around the clock and boosted both the army and air force morale. The effect of the B-17s' arrival can be understood clearly if we compare the tonnage of bombs dropped by the IDFAF during the first ten days of fighting, before and after the Flying Fortresses arrival. On the first six and half days, up to the evening of 15 July, the IDFAF dropped 9,370 Kg of bombs in 82 sorties, quite a modest 114 Kg per sortie. On the last three and half days 48,930 Kg of bombs were dropped in the same number of sorties, an average of 600 Kg per sortie, and thus the tonnage increased by more than 500 percent. Moreover, the average sortie rate per day was doubled following the arrival of the heavy bombers—from 14 to 27 sorties per day. The Ten Days fighting ended with the fourth aerial victory for 101 Squadron. That kill was important, because the previous ones (a Harvard and two C-47s) could be considered easy meat, and since the Spitfire was much superior to the troublesome S-199, and the fact that with this victory the REAF lost a senior officer, it was certainly a blow to the REAf Spit squadrons deployed at El Arish.

THE SECOND TRUCE

The Ten Days fighting ended with an United Nations-declared truce for an unrestricted period. The diplomatic efforts to end the conflict had proved to be fruitless, however, and both sides tried to achieve the best opening cards for the anticipated political solution. Israel was by then a real fact, although the Arab nations still failed to admit it. The small country had stood the Arab onslaught, and regardless of the Ten Days failures in the Egyptian and Syrian fronts, the Jerusalem corridor was finally opened following Operation Danni, and the construction of an alternative road to Jerusalem, nicknamed the "Burma Road", was accomplished, while the central Galilee, including the large Arab populated city of Nazareth, was captured during Operation Dekel. Operation Brosh, although failing to achieve its aim of forcing the Syrians to retreat from Israeli soil, was partially successful in the sense that it stopped the Syrian offensive and, until the end of the War of Independence, the Syrian front was static. Yet further to the north the Lebanese army quit the conflict as early as May, following its failure at Ramot-Nafthali, and the upper Galilee was left to local Arab forces supported in men, weapons and equipment by Syria and Iraq. Perhaps reflecting the situation then extant, the U.N. proposed to alter a little its original decision of November 1947, by giving the central Galilee to Israel and the Negev to the Arabs. Israel opposed that initiative and finally, following numerous local skirmishes, launched Operation Yoav on 15 October. This operation opened the road to the Negev, and it was the first of five operations in which the IDF defeated the Arab armies and forced their governments to the negotiation table at Rhodes. The first period of the second truce lasted from 19 July to 15 October and that timespan saw a major build-up in the IDFAF through large procurements, and yet sporadic operations were conducted as needs arose. The first such operation was Shoter (Policeman) on 24/25 July. Three besieged Arab villages blocked the traffic on the coastal road to Haifa, and the IDFAF supported the attacking forces mainly through daylight bombings on 25 July by three B-17s on the three villages, Ein Razal, Igzim and Gabba. Operation Shoter was successful, and the road to Haifa was opened.

But the main air force effort during the first period of the second truce was procurement. The Avia S-199 had proved to be a terrible fighter, although it gave 101 Squadron the ability to achieve some degree of local air superiority above central Israel. But, in order to extend it, more fighters were required. Since there was no kind of ground control, let alone radar stations, the method was to send standing patrols to the area of fighting. Such a system, in order to bear fruits, required a large number of fighters and even a larger number of pilots; the latter was no problem, but the former could not be met due to the very low serviceability rate of the Knives—in fact, only a four aircraft formation mission had been recorded on its baptism of fire on 29 May, while an attempt on 9 July to send another foursome to attack El Arish air base ended with only three, because the fourth (D-115) flown by Jacobs crashed on take-off, which was a rather common feature in the early days of 101 Squadron. The Israeli procurement delegation in Czechoslovakia complained to the Czechs about the quality of the S-199s and the not-unexpected answer was that the problem was the quality of the Israeli pilots rather than that of the fighter, and perhaps this had some truth, as 101 Sqn. pilots being highly motivated and experienced, flew very little and most of them had not flown and fought intensively since the end of WWII, and fighter's pilot capabilities are quickly lost without practicing constantly. Nevertheless, the IDFAF needed more fighters desperately, and thus the Czechs were asked for more S-199s. Then, in August came the surprising response—they had no more Avias available for export, but if the Israelis were interested, they could sell them Spitfires. This classic British fighter was a legend, and most of the volunteers who came to Israel flew it in combat during WWII. Unlike the S-199 it was a (relatively) forgiving aircraft that gave its pilot a feel of confidence, and its reliability was much higher. It is of interest to note that in the latter stages of the conflict, when the P-51D Mustang was also available to 101 Squadron, pilots like Slick Goodlin, an American volunteer who had been a test pilot back in the U.S., having gained his fame when he was the first to fly the rocket powered Bell X-1, still preferred the British fighter because it was more maneuverable and yet easier to handle, while at the same time its firepower, with two 20mm cannon and two 12.70mm machine guns, was superior to that of the Mustang, which were equipped with only four 12.70mm at the time due to a shortage of such weapons. Four Czechoslovakian squadrons served with the RAF during WWII, of which three—310, 312 and 313—were equipped with Spitfire IXs, all three squadrons with 76 Spits (known as S-89 in Czech service) flew home during August 1945 and formed the core of the reborn Czechoslovak Air Force. However, by the summer of 1948 this country was already under Soviet orbit and the Spitfires were phased out of service in preference to Russian equipment. Sam Pomerantz, an American pilot and engineer, checked the aircraft, and following a brief negotiation with the Czechs a contract was signed for 50 Spitfires for about U.S.$23,000 each. The aircraft were in need of refurbishment and the work was carried out by the Czechs at Junovice, where Sam Pomerance and Boris Senior were also detached to test fly them. At exactly the same month, August 1948, the Czechoslovakian attitude towards Israel changed due to Soviet pressure and the warm relations cooled off. One of the signs of that sudden change was the closing of the Etszion (or "Zebra" as the

A group of 101 Squadron pilots during a briefing prior to mission.(IAF)

foreign volunteers called it) airfield at Zatec and the end of the air bridge.

It was impossible to preserve secrecy in such a large operation as the Ballak flights between Israel and Czechoslovakia. On 16 July the New York Times published the full story of the air bridge under the heading "Fighter planes for Israeli army reported shuttled from Prague" and the subtitle read: "London says aircraft, acquired from Czech regime, are of Messerschmitt design—U.S. pilots man air transports." The U.S. State department exerted pressure on Czechoslovakia to end that activity on the grounds that it broke the embargo on the Middle East and the rules of the U.N.-sponsored truce. USAF aircraft overflew Czechoslovakia on reconnaissance missions and the tension grew to such an extent that on 12 August the Czechs ordered Yehuda Ben Hurin, the Israeli commander at Zatec, to evacuate the airfield by 1100 the next morning. Some fifty men and the C-46s and C-54s were immediately transferred to Tel-Nof, and the air bridge was finally closed. The unserviceable Constellation was left at Zatec with a single mechanic, Ernest Stahalic, to repair it. In less than six months some 100 Ballak flights were performed, more than one per day from the period 20 May to 12 August, with more than 600,000 Kg of equipment ferried to Israel. The load included the 25 Avia S-199s, their associated equipment, ammunition and spares. The final six Ballak flights, flown between 6 and 12 August, transported 26 tons of 70 Kg bombs for use by the S-199s and B-17s. The arrival at Tel-Nof of LATA's aircraft was a complete surprise since up to that time this airfield was merely a night stop for the heavy freighters. Moreover, this unit, so intensively flying during the previous months, was by now left without any missions, and Meir Mardor, LATA's commander, took immediate steps to reactivate the situation. The obvious step was setting up a training scheme for Israeli air and ground crews by the foreign volunteers. It was a known fact that one day the foreign volunteers would leave and the Chel Ha'avir would have to rely on its own human resources. Therefore a small number of Israeli pilots already serving with LATA were given more extensive training, Pessah Tal was the first of them to be qualified as captain when Sam Lewish checked him out on a Commando.

Meanwhile, in those days the Negev situation was desperate. It was besieged for more than three months, and although the truce agreement allowed Israel to supply it, the Egyptians blocked the road and two attempts to open it by force failed. The idea to supply this deserted area by air failed because of poor serviceability rates, lack of organization and the small capacity of available aircraft. By mid-August food stocks were down to only a week, and the Negev Brigade was in urgent need of replenishment by fresh troops from the north. Then the bulb lit up—why not use LATA's transports?. LATA (which for administrative purposes was known as 106 Sqn. in the IDFAF), compared to the air staff, was a compact body of only 20 men at the IDFAF HQ in the Yarkom Hotel, Tel-Aviv, plus another 47 at Tel-Nof. It was commanded by Meir Mardor, who had no aeronautical background but was a sound organizer, and his assistants, Leo Garner and Al Auerbach, who were highly experienced transport pilots. On 18 August Ben Gurion called a meeting to discuss the Negev situation; among others present were Meir Mardor, General Staff C.O. and Aharon Remez, IDFAF C.O. Yigal Allon, C.O. of the Southern Front claimed that his forces could break into the Negev, but they would not be able to hold an open corridor for more than ten to fifteen hours. The ball was in the hands of the airmen. To the question if the Chel Ha'avir could supply the Negev, the definitive answer was affirmative. The preparation stage

Three C-46s from 106 Squadron flying in formation over the Mediterranean. (IAF)

commenced immediately, and Leo Gardner, LATA's Operations Officer, was sent to the Negev to locate a landing strip. After examining several, he chose a landing ground at kibbutz Ro'hamma, a few kilometers east of Dorot, and work commenced to ready the place, including a dust strip of 1,120 meters length and 35 width in a roughly southwest to northeast heading. The operation was to be carried out only at night for two reasons, the first of which was to avoid interception since the flight path was over Egyptian held territory. For this purpose electrical runway lighting was installed using a 1 Kva generator, which also supplied energy to the communication set. While activity in the Negev was at full swing, meetings were held up north to define the goals of the operation and the methods to achieve them, Aharon Remez instructed LATA to deliver 600 tons during the first month of operations, which meant about 20 tons per night, while the Operations Department of the General Staff, not yet fully aware of LATA's capability, considered 15 tons per night to be quite satisfactory. LATA's staff thought otherwise and argued that they were able to deliver some 60, or even 80, tons per night to the besieged Negev. The load was defined as urgent supplies: food, ammunition, fuel and the like, plus the equipment of the Iftach Brigade which was to replace the Negev Brigade. Since the distance between Tel-Nof and Ro'hama, or merely "Avak-1" (Dust-1) as it was nickamed by the crews due to the dusty runway, was less than 100 km, the aircraft could carry the maximum payload and perform more than one trip per night. The route was above the Egyptian lines—the aircraft flying south flew at 5,000 ft, while those returning flew at 6,000 ft, and it took the Commando exactly 30 minutes to perform the flight. The climb and descent were made in circles above the airfield, either Tel-Nof or Ro'hama, while the runway lights served as references and were not turned off until the aircraft reached the planned cruising altitude or landed.

On 23 August at 18,00 hrs an empty Commando landed at Ro'hama to check the field and once it was found suitable, three other transports followed that night and discharged 29 tons of supplies. The second night witnessed eleven landings at "Avak-1" with 75 tons of supplies being unloaded. LATA's goal had been achieved. Meanwhile, number 35 Flight also came under LATA control. This unit had been organized at Tel-Nof in August with C-64 Norsemans taken from the Tel-Aviv Squadron, in addition to those which arrived from Europe under the command of Phil Marmellstein, an American volunteer. The Norsemans were used with the Dakotas of 103 Squadron at Ramat David to supply Sdom, on the southwestern shores of the Dead Sea, some 300 meters below sea level. The landing strip was only 600 meters long, and the Norsemans were flying in a straight line from Tel-Nof to the Dead Sea. After take-off they climbed above the air base to 3,000 ft at 120 mph. The safe altitude to avoid Arab AA fire over the Gosh Etszion area was 13,000 ft, but to save time and fuel most pilots flew at from 7,000 to 9,000 ft, cruising at 140 mph, being followed by a descent and a quick approach and landing at Sdom since the Jordanians fired at the aircraft during the approach leg. The carried ammunition, weapons, food and other supplies, while the Dakotas of 103 Sqn. were used to carry passengers and mail to and from Sdom, also evacuated casualties. But continuing with the air supply effort to the Negev, the record night was on 7/8 September when five serviceable aircraft made a total of 13 flights in which 50 tons were delivered. The average was 8.5 flights per night, in which 81 tons of cargo were transported to the Negev, while almost 90 passengers were flown on the return trips to Tel-Nof each night. Such an intensive operation rarely ends without accidents and incidents, and Operation Avak had its share of both. The most serious accident took place when a loaded Commando taking off from Tel-Nof suffered a failure to both engines due to a fuel system malfunction immediately after leaving the runway. The crew retracted the landing gear and performed a perfect belly landing in a field ahead. The Commando was repaired and nobody was injured. During late September the Israeli government, frustrated from fruitless diplomatic efforts to end the conflict with an agreed solution, decided to break the stalemate and take the initiative. Ben Gurion wanted to attack the Jordanians, but the government voted against his suggestion and the scene of the new offensive was to be the Negev. In anticipation of Operation Yoav, LATA was instructed to speed up its flight rate, and a second landing strip was erected near kibbutz Orim south of Ro'hama—it was known either as "Avak-2" or Imara. The higher rate of flights commenced on the night of 28/29 September, and during the following nights an average of ten sorties per night reached Ro'hama. On the night of 6/7 October a Commando suffered a technical failure at this airfield and for the first time during the operation a transport stayed in the Negev at daytime. As morning broke out, the Egyptians noticed the large aircraft and, ignoring the truce, REAF Spitfires were scrambled to attack it. Misha Kenner, the copilot that was left at Ro'hama to watch for the Commando, saw the pair of Spitfires diving at his airplane. As soon as they finished their strafing runs, Kenner rushed to the Commando to examine the damage, but to his amazement he found that not even a single round had struck it!. Four days later, on the night of 10/11 October, "Avak-2" was inaugurated by Leo Gardner with a Douglas Skymaster, to be followed during the same night by a loaded Commando with Al Auerbach at the controls. It took the Commando 40 minutes to reach Imara, compared to ten minutes less to Ro'hama. The last stage of Operation Avak was conducted during Operation Yoav, in fact it concluded on 21 October when the road to the Negev was opened, and the following statistics were recorded: in a total of 417 sorties, 2,224,235 tons of cargo and 1,911 passengers were flown from Tel-Nof to the Negev, while 271,165 tons of cargo and 3,187 passengers were flown from the Negev to Tel-Nof.

Due to U.S. pressure, the Panamian government canceled the civilian licenses on Israeli transports on 18 September, and thus for the now occasional flights aboard a dummy airline was created and appropriately named "Al-Shem", which in Hebrew means "No-Name." That title was painted on some of the Commandos and Dakotas. Another name that was used for a single VIP flight at that period was later adopted as Israel's flag carrier title: "El-Al" which means "To the Skies." That name was adopted for Operation Sahar (Dawn), which merely was a VIP flight to bring to Israel its intended president, Chaim Weizman, from Geneva. A Douglas Skymaster was chosen, and LATA's men quickly converted the freighter into a luxury transport that carried for the first time ever the title "EL-AL", and an Israeli civilian registration, 4X-ACA. On 29 September the four-engined transport picked up Weizman at

One of the American volunteer pilots which served with 101 Squadron was Rudolph "Rudy" Augarten, a WWII veteran. He also served in the postwar Chel Ha'avir, where he reached the rank of Lieutenant Colonel. (IAF)

Geneva and proceeded directly to Tel-Nof. The Skymaster crew included Al Auerbach, captain; Itschak Henenson, copilot; and Yehuda Shimoni, a Dutch volunteer, as the navigator.

Just two days before Operation Sahar, on 27 September, the second LATA C-54 Skymaster, in civilian colors and with U.S. registration (NC56011), was involved in Operation Velveta. After the end of the Ballak flights and the worldwide exposure of the links between Czechoslovakia and Israel, the former were reluctant to renew the air bridge in order to transfer the Spitfires to Israel. The Czechs insisted that in order to preserve secrecy the Spitfires should be packed in crates and transported by train to a Yugoslav harbor when an Israeli cargo ship would collect them. Such a transportation method meant that at least eight weeks would lapse until a minimum force of fighters could be organized in Israel. The radical solution, proposed by Sam Pomerance, was to fly the Spitfires to Israel either directly or with a single refueling stop. Many thought that the idea had no sense as the Spitfire was known as a short-legged fighter, even with the auxiliary lower fuselage ferry tank fitted. Pomerance's plan was to add to the Spitfires two underwing S-199 auxiliary fuel tanks, deleting guns and radios to preserve maximum take-off weight. It was estimated that such a configuration would allow the Spitfire to cover 2,250 km in about seven hours of flight—barely enough to cover the distance between Czechoslovakia and Israel with only one refueling stop. Under the guidance of Pomerance the Czechs installed the fuel system in a Spitfire and he took it for a four and half hour duration test flight. It was quite an achievement to stay in the air in a Spit for such a long time, but there was not enough time to conduct an extensive evaluation of the configuration and the real trial was to be the ferry flight to Israel.

The refueling stage was sought at first in Yugoslavia, but a negative answer was received and thus Bulgaria was approached. The Bulgarians agreed, but demanded U.S.$10,000 for each fighter on transit. Israel so urgently required the fighters that at one stage Ben Gurion instructed to pay the Bulgarians U.S.$36,000 for the passage of six Spitfires. Finally, political pressure bore its fruits (in U.S. dollars), and the Yugoslavians allowed the Spitfires a refueling stop. Former RAF B-24 pilot Gideon "Geda" Shochat was in charge of finding a suitable landing field, and found it near Pogdorice, Montenegro, close to the Albanian border in southern Yugoslavia. The distance from Kunovice to Podgorice was some 1,000 km, and from there to Israel about 2,250 km, and while the first stage presented no problems, the second posed a real challenge. It was a long overwater flight in the narrow cockpit of a single-engined fighter with a calculated fuel reserve of only twenty minutes, and thus it was decided to fly in close formation with one of LATA's C-54s leading the fighters. Once the refueling stop at Yugoslavia was organized, the planning stage commenced. It was a large scale operation by the tandards of the day and required full cooperation between the Israeli Air Force and Navy. Six Spitfires would fly in tight formation from Czechoslovakia to Yugoslavia, and following a night stop they would proceed nonstop to Israel, led by the Skymaster. The route planning was a classical maximum range problem. For propeller-driven aircraft, maximum range is achieved when the angle of attack is constant, for achieving maximum lift/drag ratio, and thus the aircraft cruise speed drops during the flight, for the Spitfire would have an initial cruise speed of 290 mph that dropped to 190 by the end of the flight. The Skymaster was to be equipped with dinghies, and if unfortunately a fighter would ditch, it would drop a dinghy to the pilot in trouble, but would continue to Israel with the rest. The Skymaster would inform about the location of the pilot in trouble to a Curtiss Commando that was to fly on a predetermined course and to the navy, two of whose ships were to be out in the open sea in predetermined points, and a third ship was at Haifa port on alert. A 103 Sqn. Dakota was also on readiness with air-sea rescue equipment at Ramat-David. Since the Spitfires were flown without guns and radios, it was necessary to organize the transportation of that equipment too, and thus a single C-64 Norseman was allocated the task of taking it from Kunovice to Podgorice, and from there one of the LATA's freighters would collect it for Israel.

On 22 September the Israeli delegation at Prague formally asked permission to fly six Spitfires to Israel. Surprisingly, permission was granted without any difficulties, and on the early morning hours of 24 September Sam Pomerance climbed into the leading Spitfire at Kunovice to prove his concept. He led five experienced pilots—which, during the last weeks had been converting or refreshing on the British fighter. They were Boris Senior, Modi Allon, Sid Cohen, Jack Cohen and Nafthali "Toxi" Blau. The flight to Yugoslavia was uneventful except for one incident. Blau was the last in the formation, and when he came in, after everybody else had landed at Podgorice, it was noticed that he had forgotten to lower his landing gear, and it was impossible to inform the pilot of his mistake as the Spit lacked radio. The accident was inevitable, the fighter was badly damaged, but Blau escaped alive. Now the real challenge was still ahead. On the morning of 27 September the formation took-off from Podgorice heading for Israel. At first all went well, then Modi Allon and a few minutes later Boris Senior discovered to their horror that their fuel gauge read Empty; the improvised fuel system had failed on their Spitfires, and both had no recourse but to divert to Rhodes, which was a better option that Cyprus, since the Greeks arrested them with charges of spying for the communists! Meanwhile, the remaining three Spitfires crossed the Israeli shores at about 1530 hrs and fifteen minutes later landed at Ramat-David air base after six hours and 45 minutes of overwater flight. Operation Velveta 1 ended with fifty percent success, as only three Spitfires had reached Israel, which would eventually be coded as D-132, D-133, and D-134.

But the operations also included a C-64 Norseman flight. On the evening of 26 September, this aircraft, piloted by Joe Sunderland, an Englishman who served with the RAF in WWII, left Kunovice. A navigation error forced Sunderland to execute an emergency landing in a field on a mountainous area some 47 km northeast of Mostar, Yugoslavia, in which the Norseman turned over on its back. He escaped unhurt, and the equipment was transferred by land to Podgorice from where it was flown to Israel by a LATA Commando. On the other hand, on 12 October an especially sent Dakota from 103 Sqn. collected Senior and Allon at Athens, after having spent 14 days in several Greek jails, and flew them to Israel. The Spitfires, however, were not released, and only after lengthy diplomatic negotiations one of the pair was returned in December 1950. The second crashed while being flown by a Greek pilot who was killed. It seems a lot of trouble for just three Spitfires, but in those days every fighter was very important.

Moddi Allon, at center, with David Ben Gurion to the left during a visit of the Prime Minister and Minister of Defense to the brand-new 101 Squadron.(IAF)

Even more troublesome and imaginative than Operation Velveta was the acquisition and ferrying to Israel of medium bombers from Great Britain—a previous attempt to purchase Spitfires had failed. Since its inception the IDFAF was keen to acquire a force of twin-engined bombers. Four A-20s were purchased in the U.S., but following the B-17 affair, were interned by the U.S. Customs Service on 25 August. However, by this date the Israelis already owned their first twin-engined fighter-bomber/attack aircraft in the shape of the British Bristol Beaufighter. During May and June 1948, Emmanuel Tsur, the former chief pilot of the then-defunct Aviron company, acquired in the UK three De Havilland Dragon Rapides, and single examples of the Miles Geminis and Aerovan. Surprisingly enough, all these light twin-engined aircraft were smuggled to Israel without any problem using a fictitious firm in which its main business was claimed to be the transport of live cancers from Tunis to southern France. Following his success Tsur was instructed to return to England and search there for suitable combat aircraft. Tsur operated from Paris through intermediate men and one of those, John Harvey, discovered two Mosquitos that were offered for sale at 4,500 Sterling Pounds each. Harvey, an ex-RAF pilot, met Tsur when he ferried for him the Miles Gemini to Israel. The Pair of Mosquitos involved were of the PR.16 (PR.XVI) photographic variant and were owned by Group Captain G.C. Chesire V.C. through a firm named VIP Association. Both Mosquitos were airworthy, and Harvey bought the pair that was registered on the name of a fictitious owner, H.L. White, another ex-RAF pilot. On 5 July John Harvey left Cambridge in the first Mosquito, G-AIRT, ex-RAF NS812. Harvey's intention was to fly to Ajaccio, refuel, and proceed non-stop to Israel. However, due to excessive fuel consumption he was forced to land at Nice, where he was arrested by the local French authorities. Following Tsur's intervention he was released and continued to Ajjacio where the malfunction that caused the abnormal fuel consumption was taken care of. Harvey stayed overnight and early the next morning left for a 2,600 km non-stop flight to Israel. Due to critical fuel shortage he decided to land at Haifa, although that airfield was not used by the IDFAF. A small British garrison was still present until 1 August, but the airfield was in Israeli hands. Not unexpectedly, the Mosquito was welcomed by small arms fire and as the "strange" aircraft stopped on the tarmac the foreign pilot was arrested by the few resident soldiers. The incident was over when Harvey explained his intentions, and on the following day, 7 July, the Mosquito arrived at Ramat David air base where it joined 103 Squadron. However, it was not a fighter-bomber and even as a photographic reconnaissance aircraft it lacked the necessary cameras and equipment and thus was little used until it was damaged on landing at Ramat-David in late September while on a training sortie, possibly to convert a pilot to the type. The aircraft was withdrawn from use for the duration of the WAR of Independence although it was repaired and re-entered service at a later date. Even less lucky was the second Mosquito PR.16, G-AIRU, ex RAF NS811.

It left England on 16 July to follow the same route of G-AIRT, but the next day it crashed during take-off from Ajaccio, allegedly due to sabotage from Arab agents, although that claim was never confirmed. The pilot escaped unhurt. The Mosquitos' procurement and smuggling was an astonishing success, but Tsur still had a few more tricks in his hat. On a roughly parallel timescale to the Mosquitos affair, Tsur discovered six Beaufighters, all of the TF.10 (TF.X) variant, at Ringway. The Beaufighters, unlike the Mosquitos, required a refurbishment, but Tsur bought them all for 1,500 Sterling Pounds each. One of the Beaufighters, G-AJMF ex RAF NV306, was used as a spares source, and towards the end of July the other five were ready for ferrying. Before attempting it, it was decided to smuggle their equipment, armament, spare parts, etc. John Harvey chartered a Handley Page Halifax C.8 (C.VIII), G-AJPJ ex RAF PP263, and left the United Kingdom bound for Israel on 20 July. Its load consisted of 41 20mm guns, 37 machine guns, spares and four engines, all for the Beaufighters. Harvey was accompanied by a former RAF armorer who was to install the armament in the Beaufighters once they arrived to Israel, and a wireless operator. At Ajaccio he received his orders from the captain of a LATA C-46 Commando and proceeded to Israel only to discover that something was totally wrong with his instructions. The lights at Tel-Nof air base had not been turned on and radio contact was not achieved, although the Halifax identified itself as "Oboe-3" as instructed. Harvey, after circling above Tel-Nof for an hour and half, low on fuel and with a hysterical wireless operator aboard, flew on two engines to nearby Sde-Dov and landed there aided by the moonlight, but the runway was too short and Harvey's former heavy bomber overran it, with the landing gear collapsing in the soft sand beyond it, but the important load was not damaged and the three men escaped unhurt, while the poor Halifax was a total loss, and was later salvaged and scrapped.

The Mosquito crash at Ajaccio, the fact that the British Consul at Haifa witnessed Harvey's landing there, and the much publicized Halifax incident meant that by then the British authorities were well aware of Tsur's activities and a special trick was required to smuggle the Beaufighters to Israel. In the preceding months Israeli agents used all kind of tricks and covers to smuggle aircraft and other military items, but it seems that Tsur's idea was the most imaginative and surprising. One of the pilots that was to ferry the Beaufighters to Israel met a pretty New Zealander girl that had been offered to participate in a film about New Zealand combat pilots in WWII. Tsur and his men immediately got the idea: let the U.S. film such a movie and once the aircraft are airborne they will disappear to the east and reappear a few days later in Israel!. Tsur founded a fictitious movie films company, R. Dickson & Partners, and a script was written about WWII operations of No 488 Squadron, RNZAF. A major setback took place during one of the test flights on 28 July when a Beaufighter, G-AJME ex RAF ND929, crashed on approach to the airfield at Thame and its pilot, M. Campbell, was killed. Nevertheless, the first day of "filming" took place as planned on 29 July, and the first scene was a scramble. After filming it, the four Beaufighters were to fly north to Scotland to continue the filming there. But, as might be expected, the aircraft never arrived. The first—and last—day of filming was over. John Harvey led the Beaufighters to Ajaccio where they were refueled, but unlike the longer ranged Mosquitos, they were unable to cover the distance to Israel nonstop and the second refueling stop was at Podgorice, Yugoslavia. The Beaufighters joined the lonely Mosquito and the Dakotas with 103 Squadron at Ramat David on 4 August as the first fighter-bombers in IDFAF service, being organized as a separate flight under the command of Leonard "Len" Fitchett, a Canadian volunteer that had been credited with the last kill of a Luftwaffe aircraft in WWII. Each of the Beaufighters received four 20mm cannon and six locally developed bomb racks under the fuselage for 250 kg bombs (two) and 70 Kg bombs (four) under the wings. One of them was used as a spares source, and of the three remaining only one was serviceable most of the days.

By 23 July IDFAF enlisted personnel strength had grown to 2,186, including 374 females, and that number reached to 5,607 by 7 October, but the number of operational air crews and skilled ground crews was quite low, and for a proposed force of three fighter squadrons, either more foreign volunteers would have to be recruited or young Israelis trained as fighter pilots, and only five years later this goal would be achieved. In fact, during the immediate postwar period in 1949, 101 Squadron was still the only fighter unit within the IDFAF—the limiting parameter was aircraft, not skilled crews. An empty start in this aspect was when the Spitfires were acquired in Czechoslovakia, since when the fifty Spits were purchased it was planned to activate a second fighter outfit, 105 Squadron, led by Boris Senior, while several of the pilots posted to it were sent to Czechoslovakia to ferry them in Operation Velveta 1. But after only three of the intended six fighters arrived safely to Israel, the plan to form 105 Squadron had been abandoned, and this unit would not see the light until more Spitfires were received and more pilots trained, as well as when 101 Sqn. started its full reequippment with P-51 Mustangs in December 1950.

Meanwhile, by 29 July 1968 the Chel Ha'avir had a strength of ten Avia S-199s, a small number of Beaufighters and Mosquitos, three B-17s, three Dakotas, one DC-5, seven C-46s, one C-54 and 18 light-planes. By 17 September that force had grown as follows:

TYPE	TOTAL NO.	IN USE	IN REPAIR
Avia S-199	13	9	4
Spitfire 9	2	1	1 (1)
P-51D Mustang	2	0	2
B-17G Fortress	3	2	1
Beaufighter TF.10	3	2	1 (2)
Mosquito PR.16	1	1	0
C-46 Commando	7	5	2
C-54 Skymaster	2	2	0
C-69 Constellation	1	0	1 (3)
Lodestar	1	1	0
C-47 Dakota	2	2	0
DC-5	1	1	0
C-64 Norseman	5	3	2
DH Dragon Rapide	5	4	1
Fairchild Argus	2	0	2
RWD.13	2	1	1

Auster AOP	9	4	5
Lockheed Seabee	1	1	0
Vultee BT-13	1	1	0
Grumman Widgeon	2	2	0 (4)
Nord 1203	2	2	0 (5)
Piper Cub	20	2	18
Taylorcraft	2	0	2
Beechcraft Bonanza	3	0	3
TOTAL	92	48	43

(1) - Assembled from wrecked RAF and Egyptian airframes.
(2) - The one reduced to spares was w/o from the inventory.
(3) - At Zatec, Czechoslovakia.
(4) - Donated by Dutchman Bernard Van Leer.
(5) - Donated by a Jewish family from France and configured as an aerial ambulance.

Operation Shoter was mentioned earlier in this narrative, but it was not the only operational activity during the second truce. In that period the IDFAF lost three light aircraft, the first two during routine activity and the third on a photographic operational sortie. On 3 August Emmanuel Rotstein of the Tel-Aviv Sqn. was sent in an Auster to collect Zohara Levitov, the only female pilot in the air force during the War of Independence, from her parents' house where she spent her holidays. The landing strip at Jerusalem was little used by then because of the reopened Jerusalem Corridor and the newly constructed "Burma Road", and a few weeks later it was abandoned completely. After the Auster took off from the hill-surrounded landing strip it struck one of them and crashed, and its two occupants were killed. Another Auster crashed on 19 September, this time without casualties. Earlier, on 14 August, Paltiel Maklef, pilot, and Monti Goldberg, photographer, were sent in a Tel-Aviv Sqn. Fairchild Argus to photograph Egyptian positions in northern Negev. To send such a slow and vulnerable aircraft on a demanding and dangerous task seems almost a suicide. Moreover, the lonely Argus was not provided fighter cover and it seems that no CAP was present, although its presence would not have changed the Argus' fate. If it is considered that fighters were available, then it should be concluded that planning was bad by any standards. The Argus flew from east to west over the Dead Sea, past the northern Negev and to Gaza. Maklef was flying while Goldberg was photographing with his hand-held camera. Over Gaza they encountered heavy AA fire and the Argus was damaged, but it was still flyable, and Maklef turned for home, but then it was intercepted by a pair of REAF Spitfires and, after suffering more damage, crash landed behind Egyptian lines following the unequal combat. The Argus was wrecked and the crew was captured by Egyptian forces.

The score was evened more than one month later, although it was also an unequal engagement and one to not be proud of. On 22 September Gideon Lichtman was flying a patrol in a S-199 when he discovered a Dragon Rapide in red livery with Arabic titling overflying Sfat in the central Galilee at an altitude of 2,000 ft and on a northwestern heading. Lichtman instructed the Dragon Rapide to follow him by lowering his landing gear, but either the pilot (a Britisher) did not understand, or he decided to evade. The twin-engined biplane made a 180-degree turn, so Lichtman fired a short warning burst in front of him. The Rapide pilot kept ignoring the Israeli fighter, so Gideon fired a second burst that set the Dragon Rapide on fire, the aircraft crashed into Jordanian territory, and the pilot and two passengers survived, while another three, including two British journalists, were killed. The Dragon Rapide belonged to the Jordanian company Arab Airways Association, and its registration was TJ-AAQ.

Previously, between 6 and 16 September, 101 Squadron had a spate of bad luck with four of its troublesome Avia S-199s damaged or destroyed in accidents, D-115, D-120, D-121 and D-122, of which 120 and 121 were repaired. Thus, on 15 October 1948, the eve of Operation Yoav, the IDF offensive planned to recover the Negev from the Egyptian army, 101 Squadron had eight S-199s and four Spitfires available, and with this meager force would gain air superiority over the battlefield.

Two 101 Squadron pilots, in different flying garb, pose beside a Knife, as the S199 was nicknamed in the Israeli Air Force. On the left is Lee Sinclair and on the right, Buck Feldman. (IAF)

Chapter 6: Victory

On 15 October 1948, Israel launched Operation Yoav against the Egyptian forces in northern Negev. It was a political decision of the Israeli government to break the deadlock and force the Arab nations to the negotiation table by a military victory over the leading Arab country, Egypt. As it will be seen in the following lines, Operation Yoav did not achieve that goal, but it was the first in the succession of five major operations, and a few others of less significance, that were to put an end to the struggle. However, from a pure military point of view Yoav did achieve its intended aim, which was to open the road to the besieged Negev. For that purpose three IDF infantry brigades, the Negev Brigade, the Iftach Brigade and the Givatti Brigade, were massed. The forces enjoyed the support of an armored regiment (a pair of British Cromwells and a few French Hotchkiss), the Artillery Corps, the IDFAF, and the Israeli Navy. It was the first large-scale operation with such cooperation among the various arms of the IDF, and just like in the Ten Days fighting the plan was to open the operation with a dusk attack on El Arish air base.

By 15 October, however, the IDFAF had grown in human and material capabilities and its ORBAT for this date shows that 101 Squadron at Hertzlia had mixed equipment in the form of seven combat ready Avia S-199s and four Spitfires also combat ready, one of each capable of photo recce operations, with the squadron being led by Modi Allon. At Ramat David 103 Squadron had five serviceable Dakotas and a flight of Beaufighters with two serviceable machines under the command of Danni Rosin, and 69 Squadron with three B-17s, all serviceable, led by Bill Katz. At LATA's Tel-Nof air base was the 106 Squadron with six C-46s, a pair of C-54s and a single Lodestar was resident alongside 35 Flight with a small number of Norsemans. The three sectorial squadrons operated an assortment of light aircraft on their regular missions. Opposing that force, the REAF still had numerical superiority in the fighter arena since it operated two Spitfire squadrons from El Arish. It was planned that the IDFAF should perform three types of missions during Operation Yoav in addition to their regular tasks: Combat Air Patrol above the front lines, Close Air Support and preplanned attacks on important Egyptian installations, such as the El Arish air base and HQ and barracks at Gaza. All these were executed under a separate operation named Egrof (Fist) in which a constant pressure was to be exerted on the Egyptian forces by a round the clock, day and night activity. The original plan, which worked quite well, for a pleasant change, was to use all the IDFAF units. The original Operation Egrof planned attacks for 15 October included three simultaneous opening attacks at 1740 hrs on El Arish air base and the Egyptian Army HQ at Gaza and at Majdal, where a forward brigade HQ was located. Four Spitfires and two Beaufighters were to attack El Arish. Four C-46s, three Dakotas and four S-199s were to raid Gaza, and the three B-17s were to bomb Majdal. Following that dusk operation, a second wave of night bombing was to follow

RCAF pilot John H. McElroy flew Spitfires and Mustangs with 101 Squadron, being involved in the 7 January 1949 dogfight with RAF fighters.(RCAF)

with the three Fortresses bombing Gaza, El Arish and Majdal at 2300 hrs, and a single Dakota bombing Be'er Sheva at 2330 hrs, and thus ending the first day of Operation Egrof. For the two types of aircraft which opened Operation Yoav with the El Arish raid, it was their operational debut with the IDFAF. A pair of 103 Sqn. Beaufighters that took off from Ramat David were escorted by three 101 Sqn. Spitfires that took off from Hertzlia, and the fourth went unserviceable. At 1740 hrs the five Israeli aircraft were above their assigned target and the first effective attack against this REAF airfield was launched. Its defenders were completely surprised and not a single interceptor was scrambled to meet the raiders. Even the massive AA array defending the base—according to Israeli intelligence there were some twenty guns—did not open fire during the initial stage of the attack. The Beaufighters dropped from very low level their load of two 250-kg and four smaller bombs each and then strafed, together with the Spitfires, the parked Egyptian Spitfires. The attackers claimed four Spitfires either damaged or destroyed on the ground and a direct hit in a large hangar was also recorded. However, once the AA defenses went into action they proved to be quite dangerous and one of the Beaufighters was damaged and was forced to perform an emergency landing at Tel-Nof on one engine. The attack was a success and El Arish was temporarily closed.

However, one of the other opening attacks was a relative failure, while the other was a complete failure, and both were executed hours after the planned time over target of 1740 hrs. Of the three 103 Sqn. Dakotas intended for the Gaza raid, only two were loaded with bombs by dusk and even that pair left Ramat-David only as darkness fell. As usual, one of the four S-199s went u/s and thus only the four 106 Sqn. Commandos were escorted by the three Knives led by Modi Allon, either because of the Dakota's lower cruising speed or because of their late departure. However, the C-46s completely missed their targets, while the Dakotas that lagged behind bombed Gaza harbor. The Egyptian expeditionary force stationed in northern Negev and northeast Sinai received its supplies from mainland Egypt through two main channels: the train from Port Said to Gaza (each day a train with some 30 freight cars arrived to Gaza with supplies) and by ships to Gaza harbor, and both kinds of supply lines received full attention by the IDFAF and the Israeli Navy during Operation Yoav. Naturally, because of their late arrival the Dakotas encountered heavy flak, but returned unscathed to Ramat-David. It was the operational debut of the electrically operated external bomb racks on the Dakotas, while the Commando bombing method was still the "manually" door-dropped 50-kg bombs. These were passed from hand to hand within the fuselage by a chain of young air cadets. The total bomb load dropped on Gaza by the six makeshift bombers was five and half tons. While this raid was a relative failure, the third and final mission on the evening of 15 October was a complete failure. The three B-17s of 69 Squadron were planned to bomb the Egyptian HQ at Majdal, north of Gaza, at 1740 hrs exactly when the El Arish air base was successfully attacked. Like their neighboring 103 Squadron at Ramat-David, the B-17s took off late and, moreover, instead of bombing Majdal, dropped the 4500-kg bomb load on a nearby village. As a consequence the plan to bomb Majdal, El Arish and Gaza at 2300 by a single B-17 each had to be delayed to 0115 hrs, 16 October, and once more the Gaza raid failed, the target was not located, and the bombs dropped into the Mediterranean, although the other two B-17s attacked Majdal and El Arish successfully. Also, at around 0100 hrs instead of 2330 as planned, a single Dakota bombed the Egyptian forces at Be'er Sheva in the central Negev. This poor showing could still be attributed to poor discipline and organization, which perhaps is explainable due to the IDFAF youth and that it started to work as an air force in a shooting war, although the El Arish raid revealed that there were some improvements.

During the same night, 15/16 October, the Iftach and Givatti Brigades attacked the Egyptian forces in northern Negev and blocked two key roads, Majdal-Gaza and Majdal-Beit Gobrin. The next morning the main effort was launched when the Negev Brigade, supported by the armored regiment, attacked Iraq El Manshia. According to the original Operation Egrof plan, a pair of Spitfires was to stay all day on readiness at Hertzlia in case the REAF would retaliate with air raids on Tel-Aviv. The available S-199s were to mount a continuous CAP over Falluja in northern Negev to protect the Israeli forces fighting at Ira El Manshua. The rest of the IDFAF available inventory was to execute the following missions:

0625 hrs: two Beafighters to Falluja.
0900 hrs: three B-17s escorted by two Spitfires to El Arish AB
1330 hrs: three B-17s escorted by two Spitfires to Majdal.
1330 hrs: two Beaufighters to Falluja.
1815 hrs: three B-17s to Gaza.

As might be expected, some alterations were made to compensate for the ground forces demands and the availability of the aircraft. Since one of the two available Beaufighters was damaged the previous day, it was decided to cancel the planned Beaufighter missions. At 0700 hrs a single B-17 bombed Majdal, and at 0930 the three heavy bombers left Ramat-David to bomb El Arish air base, but the planned escort of two Spitfires was absent. The only appearance of REAF's Spitfires that day was about 0700 hrs when a pair of patrolling S-199s intercepted two hostile Spitfires, and one was claimed damaged. Murray Mann, a member of the next S-199 CAP, force-landed near Petah Tikva on his way back to Hertzlia due to mechanical failure. The Avia was completely wrecked, but Mann, one of 101 Sqn. flight commanders, escaped the incident alive, although it seems that his flying career with the IDFAF was over due to the injuries he suffered during the crash landing. Following the absence of REAF fighters, the S-199s on CAP gave close air support to the IDF forces fighting at Iraq El Manshia. By noon the attack on this strongpoint had failed, and the IDFAF was called to support the retreating forces. The only available Beaufighter was sent, and later on, at 1600 hrs, the three B-17s, escorted by a single Spitfire, bombed Egyptian artillery at Falluja. While IDF forces retreated from Iraq El Manshia, the Egyptian forces at Ashod were retreating south to Majdal and a pair of S-199s were sent to attack them. Modi Allon led Ezer Weizman to bomb the Egyptian forces with a pair of 70-kg bombs each, plus several strafing runs. On their return to Hertzlia at 1735, Modi Allon crashed while on finals and was killed. The cause of the crash was either previous

The 7 January 1949 combat ended with several RAF Spitfires and Tempests lost to IDFAF Spitfires and small arms fire.(IAF)

battle damage or loss of control. It was a major blow to the young squadron, which on a single day lost its beloved commander and one of its two flight commanders. The South African Sid Cohen was immediately appointed as commander, while the American Rudy Augarten and the Israeli Ezer Weizman were his respective flight commanders.

The first day of fighting ended with mixed results: the Egyptian forces at Iraq El Manshia held on against the IDF attack, while the Israeli forces' threat on the Majdal-Gaza road forced the Egyptian forces to retreat to the south from Ashod. That night the Southern Front commander, Yig'alAllon, decided to abandon the effort at Iraq El Manshia, east of Falluja, and attack the Egyptian forces occupying several strongpoints, locally known as the Junction Strongpoints, on the main road to the Negev just west of Falluja and east of Majdal. Also on the night of 16/17 October, Israeli forces penetrated deep into Egyptian territory and sabotaged the Port Said-Gaza railroad between Rafah and El Arish and between Rafah and Khan Yunis. While the Givatti Brigade launched its night attack, the Chel Ha'avir was busy with its own night missions. 106 Squadron still maintained the air link to Rohamaas part of Operation Avak, bringing in supplies and evacuating casualties. During the following nights, of the five Commandos and one Skymaster available to the squadron, three of the former executed night bombing missions over Gaza, while the rest flew to Rohama. At 0300, 17 October, two Commandos left Tel-Nof for Gaza, which was located only when Egyptian searchlights were turned on to direct the flak. In one of these aircraft flew Aharon Remez, IDFAF commander, Heyman Shamir, vice-commander and Meir Mardor, LATA's commander, to get a first-hand impression of that night bombing . Had the Egyptians downed that particular Commando, the high command of the IDFAF would have crashed instantly. The C-46s encountered heavy but inaccurate flak in spite of the searchlights and the long time over the target required to drop the full bomb load in the manual method. Meanwhile, the Dakotas of 103 Squadron bombed that night El Arish air base, Falluja and Majdal. The night attack performed by the Givatti Brigade was successful, and on 17 October the Oded Brigade was transferred to the Negev from the Galilee to augment the fighting forces and exploit to the full the local success of the Givatti Brigade. However, the Egyptian forces retaliated with several counterattacks on the Junction Strongpoints on 17 October, and even more importantly the main road to the Negev was still blocked due to the presence of Egyptian forces at Holikat, south of the Junction Strongpoints. Unlike the previous day, REAF's Spitfires appeared that day in force to support the hard-pressed Egyptian ground forces, and it was suspected that the REAF Spitfires operated from an emergency landing strip south of El Arish, Egyptian Spits appeared in relatively large formations of four, even six aircraft each, and the IDFAF failed to preserve its total air superiority achieved on the previous day, perhaps as a result of Modi Allon's death. The only Israeli aircraft which operated on 17 October were the B-17s that bombed Falluja during the afternoon. The night of 17/18 October witnessed the usual activity by 103 and 106 Squadrons, but all the IDF ground forces attacks failed—it seemed that while the REAF was reappearing in the Negev skies, the Israeli air force and army were losing their opening momentum: against the 60 and 54 operational sorties executed on 16 and 17 October respectively, only 34 were reached on 18 October, with Egyptian Spitfires attacking that day several IDF targets including Negba, Rohama and Dorot.

Following the Commando and Dakota air raids during the night of 18/19 October the number of operational sorties grew to 52, still below the 16 October record, but a marked improvement to the previous day. It was on that day that a successful air-sea cooperation was achieved for the first time. The Israeli Navy task during

Operation Yoav was to deny the Egyptian Army its seagoing supplies and to shell coastal strongpoints. On 19 October three Israeli ships located an Egyptian supply vessel north of Gaza and opened fire. The Egyptian called for air support and REAF Spitfires appeared and attacked the Israeli ships. One seaman was killed and four others injured, included Mocca Limon, the C-54 navigator on Operation Hassida and the Israeli Navy Ship Haggana's executive officer. The IDFAF was called to support the hard-pressed ships. A single Beaufighter, flown by the Canadian Len Flichett, was sent to attack the Egyptian ship, but it was intercepted by a REAF Fury. Fitchett, well aware of the superior performance of the hostile fighter, jettisoned his bomb load to the sea and dived for the water's surface closely pursued by the Egyptian. At the last moment Fitchett pulled up and recovered from the dive, but his opponent did not and the Fury disappeared into the water. The IDFAF claimed that it was an IrAF Fury, but it seems more likely that it was the Fury demonstrator, G-AKRY.

That day the B-17s bombed Gaza and Holikat, and on the last-mentioned target were escorted by Spitfires. One of the Spitfire pilots was Joseph John Doyle, another Canadian volunteer on his first operational sortie with 101 Squadron. Doyle, who was to become one of the most successful fighter pilots of the War of Independence, reported that the bombing was accurate and that Egyptian flak was moderate. It was on the next day, 20 October, that the main phase of Operation Yoav was to end with the opening of the road to the Negev. During the night of 19/20 October the Givatti Brigade finally overtook Holikat after a fierce battle that closed down to bayonet range, losing some 25 of its men, with another 70 injured, but the Egyptians left behind some 70 bodies and another 50 were taken POW. During that important night the IDFAF made its greatest effort. Norsemans from 35 Flight bombed Be'er Sheva, the light-planes of the Negev and Tel-Aviv Sqns. bombed Dimara and Yad Mordechai, and the Dakotas of 103 Squadron bombed Gaza, Majdal, and El Arish air base, in the latter case at 2340 and 0015 hrs respectively. That night was of significance because the road to the Negev was finally opened and the Egyptian forces were soon to withdraw towards Gaza, leaving Negev to Israel. However, it was on 20 October when yet another attempt to widen the newly-opened corridor by attacking the ex-British Tegart fortress at Iraq El Suidan failed with the loss of Canadian volunteer Len Fitchette and his crew. It was planned to attack Iraq El Suidan the previous day, but it was postponed since the pilots who should take part had to train in low level attack tactics for such a small and heavily defended target. The aerial attack that was to proceed the ground forces attack on the fortress consisted of a pair of 103 Squadron Beaufighters. Leading them was Len Fitchette, the commander of the Beaufighter Flight, with the ex-RAF navigator Dov Shugerman and Stanley Andrew, an ex-USAAF WWII pilot in the Pacific Theater that joined the two-man crew in spite of orders to the contrary. Fitchette's wingman was Danni Rosin, 103 Sqn. C.O. Fitchette's brief was to make a single, very low level bombing run, but on the first pass he did not drop all his bombs, while Rosin's bombload refused to leave his plane due to mechanical failure. Fitchette ignored his own instructions and returned for a second pass from a different heading with Rosin, who in spite of the mechanical malfunction closely followed him. This time the Egyptian defenders were ready, and the leading Beaufighter received a direct hit in the left engine and crashed behind Egyptian lines, and all aboard were killed. The ground forces attack failed and was forced to retreat due to superior Egyptian firepower. However, on the evening of 20 October it became clear that the Egyptian front was collapsing. At 2000 hrs an intercepted message was delivered to David Ben Gurion in which the local Egyptian commanders requested from their superiors at Cairo an approval to retreat from Ashdod, Majdal and Iraq El Suida to a new line of roughly Gaza to Be'er Sheva, and later on that evening the positive Egyptian answer was also intercepted. The local Egyptian commanders were approved to retreat immediately. The IDF had two options open to take advantage of the Egyptian army plans; a thrust towards Gaza or an attack on Be'er Sheva. The latter option was taken, and the place was captured on 21 October. On that day Rudy Augarten led Joseph Doyle in a pair of Spitfires on a CAP mission between Be'er Sheva and El Arish. The pair intercepted a formation of four REAF Spitfires and claimed two shot down—the victor was Augarten—while damaging a third; clearly 101 Squadron was starting to recover from the shock of Moddi Allon's death less than a week earlier.

While the Israeli forces achieved the initial goals set for Operation Yoav, the Egyptian forces were slowly retreating from Majdal and Ashdod towards Gaza. The Egyptian 9th Brigade, under the command of the Sudanese Brigadier Said Taha, did not retreat from its positions at Fallujam Iraq El Manshia and Iraq el Suidan and became besieged in the so-called "Falluja Pocket" by the end of October. During Operation Yoav, the IDFAF made a total of 239 sorties, which was an average of about 35 per day. This total included 34 daytime bombing sorties by the B-17s, 31 nocturnal bombing sorties by the Commandos, and a total of 85 sorties by the

Close-up of the tailfin of a 69 Squadron B-17G with an unnoficial insignia.(IAF)

light-planes of the Tel-Aviv and Negev Sqns. The peak was 16 October, the first full day of fighting, in which 60 sorties were carried out. The total quantity of bombs dropped by the IDFAF during Operation Yoav was 151 tons—an average of 22 tons per day—and, although the actual damage caused was quite light, the psychological effect, both on the Egyptian forces and the IDF units, was highly appreciated. During the week or so of Yoav the Israelis lost three aircraft, with the loss of three highly experienced pilots, two killed and the other wounded. To these losses another one, although not directly connected to Yoav, should be added. On 25 October a Dakota crashed at Tel-Nof air base and all its six occupants were killed. The crew consisted of four; the captain was Wolf "Zeev" Canter, a Canadian volunteer that served as a bomber pilot with the RCAF in Europe during WWII. The other three crewmembers were the copilot Fred Stevenson, the navigator, William Fisher, and the wireless operator, Yehuda Litman. One of the two passengers was Michael Weimers, who contributed so much to operation Avak. Against these losses the IDFAF claimed three REAF destroyed in the air and another five on the ground at El Arish air base.

With an open corridor to the Negev and the Egyptian army losing its initiative, the Northern Front was urging a similar thrust to capture the Central Upper Galilee, which was under the control of a volunteer force organized by the Arab League even before the establishment of Israel. That force consisted of several thousands under the command of Lebanese Colonel Fausi El Kaukgi. The offensive was named Operation Chiram, and while in the south the IDF faced a regular army, Kaukgi's forces were mostly irregulars and lacked any form of air support, although Iraqi aircraft were present during some of its attacks on Israeli settlements during the early stages of the War. Thus, the IDFAF was less committed to CAP and concentrated on support missions. Operation Chiram was opened on the night of 28/29 October with four brigades participating, including the forces that held the positions against the Syrian bridgehead at Mishmar Ha'yarden and the Iraqi forces in the Lower Galilee, to secure the attacking forces from any unexpected surprises. Even before the ground forces action began, and much like the opening IDFAF attack in Operation Yoav, the three 69 Sqn. B-17s, escorted by 101 Sqn. Spitfires, were sent to bomb Kaukgi's HQ

at Tarsshicha. However, because of the similarity of the terrain and villages in the area and the relative lack of acquaintance with the Galilee by the heavy bomber crews that were accustomed to operations in the Negev, the B-17s bombed the village of Sochtama by mistake, and the Oded Brigade failed to capture Tarshicha. On the morning of 29 October, between 0700 and 0745, Tarshicha was finally bombed by two B-17s, again escorted by Spitfires, although aerial opposition seemed very unlikely in that area. This second attempt enabled the Oded Brigade to capture the town from the west with the consequence that Kaukgi and his main force immediately retreated north to Lebanon. Since the 7th Brigade thrust towards Tarshich from Sfat in the east and the Gollani Brigade attack from the south were also successful, the Central Upper Galilee was completely captured by 31 October, as well as 14 Lebanese villages west of the Chula valley and as far north as the Litani river. During the short and successful Operation Chiram, in which it performed mainly in the support role, the IDFAF recorded a total of 34 sorties—an average of almost 14 per day—dropping 27 tons of bombs—almost eleven per day. Following both operations, Israel's position was greatly improved and it had no besieged areas except for Sdom on the southeast corner of the Dead Sea. The Arab nations had by then lost their initial enthusiasm and unity. The Lebanese were long out of the conflict, the Syrians were holding their positions without any attempt to launch an offensive, the Iraqis failed to cut off the Galilee from central Israel in the Sharon area, and the Jordanians were conducting secret talks with Israel in maintaining the status quo in the Jerusalem area while the Egyptians retreated towards Gaza, leaving behind their 9th Brigade besieged at Falluja. Yet the Arab nations refused to admit the real facts, and under British pressure the U.N. called on 4 November all sides to retreat to the 14 October positions; naturally, Israel refused to lose such an advantage and called the Arab nations to the negotiations table. While the U.S. was exerting pressure on Egypt to open cease-fire talks with Israel, the British were playing the wrong cards, a policy that was to cost the RAF six aircraft soon. The UK had mutual agreements with Jordan and Egypt, and large British forces were present in the Middle East, especially in the Suez Canal Zone. Following the Israeli successes in Operation Yoav it seems that the local British commanders overestimated the IDF capabilities and they feared of an attempt to outflank the Egyptian forces at Gaza by a landing from the sea at the area of the northern entrance of the Suez Canal. To keep an eye on Israel's activities the RAF was instructed to execute reconnaissance missions, and the regular intruder became known to IDFAF personnel as Noon Charlie. That aircraft (it was unknown at that stage that it was British) entered Israel's airspace in the Galilee from Jordan, overflew Ramat-David air base, and continued south to the northern Negev before leaving Israel into the Sinai peninsula. This airplane was in fact an RAF Mosquito PR.34 from No 13 Squadron based at Kabrit in the Suez Canal Zone. It flew at an altitude of about 30,000 ft, and its white vapor trails were seen for miles in the clear blue autumn skies of Israel. However, the RAF fliers were so confident that they made no attempt to conceal themselves or even change their course from time to time; as long as the IDFAF was equipped with oxygenless Avia S-199s, they indeed had nothing to fear. Moreover, a successful interception was only possible if sufficient warning was given to the fighter. On the other hand, the Spitfires had oxygen and an intercept attempt was made on 22 October by Joseph Doyle, who took off from Hertzlia but could not intercept what he thought to be an "Iraqi Mosquito." Meanwhile, the supplies of aircraft from the U.S. improved as they could be smuggled by sea as spares and agricultural equipment. On Saturday, 17 July, the cargo ship Enterprise left New York harbor carrying a pair of Piper Cubs, a pair of Mustangs and 18 Pratt & Whitney R-2800 spare engines for the C-46s. A report from 19 August confirms that the Enterprise offloaded its cargo in Israel by that date and the two Mustangs were transported to the CMU (Central Maintenance Unit) at Mabarot for assembly. An order of battle report from 17 September claims that the Mustangs were still unserviceable. The long period of assembly could be attributed to several factors: the packing of the Mustangs in crates was by unskilled personnel whose main interest was smuggling, and as a consequence the aircraft were damaged, while the assembly stage by the CMU was without technical manuals and its staff was not familiar with the type—even some vital pieces of equipment were missing, such as the gunsight, and thus, in a makeshift way, the Revi gunsights from wrecked S-199s were installed in the P-51s. Therefore, the two American fighters were never delivered to 101 Squadron while based at Hertzlia, but after returning to Hatzor air base. Following Operation Yoav and the retreating of the Egyptian forces, the front line moved further away from Hertzlia, while at the same time the Egyptian artillery no longer threatened the ex-RAF station of Hatzor.

Hatzor was constructed during WWII and it had most of the facilities intact, including concrete runways, control tower and hangars. Thus, on the second week of November 1948, 101 moved from Hertzlia to Hatzor with its S-199s and Spitfires, along with the squadron hacks, a Republic Seabee, a Vultee BT-13 and a Piper Cub. This move was to prove fatal as far as Noon Charlie was concerned. While the distance between Ramat-David and Hertzlia is less than 100 km, the distance between Ramat-David and Hatzor is almost 200 km, and it was that margin, coupled with the arrival of the faster climbing Mustangs, that enabled the elimination of Noon Charlie on 20 November. Seven days before the last scramble from muddy Hertzlia was made, American volunteer Charles "Red" Nott, aboard Spitfire D-132, tried to shoot down the recce Mosquito, albeit without success. On the late morning of 20 November, the control tower at Ramat David announced to Hatzor, by a telephone call, that Noon Charlie had arrived. The longer distance between Ramat-David and 101 Sqn. base improved the available climbing time by almost 100 percent. It was the American volunteer Wayne Peake that rushed to the waiting Mustang while squadron commander Syd Cohen climbed to the control tower and grabbed a pair of binoculars. Cohen closely tracked the hostile aircraft and directed Peake through the radio to his prey. The Mustang pilot almost fainted due to a defective oxygen supply, but aided by Cohen's instructions he got in behind the Mosquito, which probably noted him since it turned west towards the Mediterranean Sea instead of continuing towards the Sinai as usual. Peake fired a long burst, and the Mosquito exploded in the air, the debris falling into the sea. Peake regained full consciousness and landed at Hatzor, thinking that he had downed a Halifax!. Upon observing the outcome of the interception from the ground, Ezer Weizman took off in the amphibious Sea Bee to the scene of the crash, off Dor's beach, to look for survivors, but he could only observe some of the remains of the aircraft floating on the water. Only after the 7 January clash between the RAF and the IDFAF did the British government admit the loss of that No 13 Sqn. Mosquito. Its unfortunate crew, P/O Reynolds and F/Sgt Lowe, were never found. During November, IDFAF activity concentrated in the Falluja pocket area. Its goals were to deprive the besieged Egyptian brigade from aerial supply by the REAF, and to bother them by day and night bombing, thus destroying or minimizing their will to fight. On 30 October it was reported that seven REAF Spitfires overflew the pocket—possibly to drop supplies—and Brigadier Taha wrote in his diary that six REAF aircraft dropped supplies at about 1000 hrs that day, but later had to stop

Two B17Gs of 69 Squadron during an operational sortie; note bombing mission markings on the nose of the nearest Flying Fortress. (IAF)

that effort because of the presence of Israeli aircraft in the area. That effort continued on 31 October and ground forces claimed one REAF aircraft downed. On 5 November the IDFAF commenced its bombing offensive on Falluga Pocket, and in the following weeks a pattern emerged; the three B-17s bombed at dusk, while the transports of 103 and 106 Sqn.s and the light-planes of the Negev Sqn. bombed at night. On 9 November the Iraq El Suidan Teggart fortress was finally captured, and the pocket was reduced. The three B-17s supported the successful attack of the 8th Brigade by bombing the HQ and artillery batteries of the Egyptian brigade at Falluja. On 11 November, and already operating from Hatzor—in advance of the rest of the squadron, which was still at Hertzlia—Rudy Augarten and Boris Senior, flying Spitfires, intercepted a REAF Dakota and shot it down. The Chel Ha'avir goal of eliminating the besieged brigade's will to fight was not achieved by those bombings, which caused little damage. The accuracy of the night bombing transports was especially poor. For example, on the evening hours of 17 November a pair of Norsemans bombed Hatzor air base instead of the Falluja pocket. On 18 November it was decided to stop the bombing as it was estimated that the number of casualties among the well-dug Egyptians did not justify the effort. One day before, 101 Squadron Spitfires intercepted three REAF Spitfires at 7,500 ft over Falluja pocket, and Rudy Augarten claimed one shot down. The growing confidence within the IDFAF during the autumn of 1948 led to a series of reconnaissance missions in which the Spitfire number D-130, built from scrap, played a key role. During October, D-130 was used to photo El Arish air base, the Egyptian strongpoints in the Negev, and Jordanian positions in the Latrum-Ramallah area north of the Jerusalem corridor. Because of the short focal length of the vertical-mounted camera lenses and the lack of a reliable oxygen supply system, which restricted altitude to 14,000 ft, the field covered by the camera in each run was quite narrow, and thus many runs were required to properly cover the target, which meant spending at least half an hour above it. Moreover, in order to get good quality photos the airspeed was restricted to 220 mph, and those two factors, long time above the target and relatively low speed, made the recce Spitfire an easy target. Thus, for dangerous sorties it was given an escort which usually consisted of a single fighter—at first another Spitfire, and later on a Mustang. The recce Spit, although built from scratch by the CMU, had a very good reputation among 101 Sqn. pilots and only occasionally suffered from minor technical failures such as the jamming of the film in the camera, which caused several missions to be scrubbed. The first long range, deep penetration reconnaissance mission in the IDFAF history was performed in mid-November. On the 19th, Ezer Weizmann, in the recce Spitfire, and Charles "Red" Nott, in an escorting Mustang, left Hatzor for Syria. They refueled at Ramat-David and took off at 1150 hrs for their two assigned targets, Damascus and Sach El Sacher. Weizman flew steady at 14,000 ft and 220 mph and made three runs over Damascus, which took some 40 minutes, and another three runs over Sach El Sacher. The pair was not disturbed by flak of any kind, and the SAF, equipped with Harvards, made no attempt to intercept. Weizman and Nott landed at Ramat David at 1439 hrs, and the photos they brought back were excellent. In one of them of El Maza airfield in the outskirts of Damascus, two C-47s, two T-6s and a single-in line engined fighter were clearly visible. Later on Joseph Doyle participated in a similar mission to Damascus and Beirut. In early December the attention was turned to Jordan, and the reconnaissance Spitfire visited Amman and Mafraq, where Iraqi aircraft were based. Avro Ansons and Douglas Dakotas were observed in the airfield, but none of the feared Hawker Furies.

Sometimes success invites trouble; Operation Yoav and a lesser known one named Operation Lot, which opened the road to Sdom on 23 November, forced the transport squadrons to reduce activity to almost nil. The opening of the Negev road ceased Operation Avak, and thus freed 106 Sqn., while the opening of the Sdom road relieved 103 Sqn. and 35 Flight from their obligations to supply that remote area. Those units, and especially 106 Sqn., consisted almost exclusively of foreign volunteers. As long as activity slowed down and their spare time grew, the troubles began. The atmosphere was that the crucial stage of the struggle to the survival of Israel was over. To the foreign volunteers of LATA, it was more than enough. It is true that some of them came for the money, but most, unlike the fighter pilots that were mostly adventurers, were Jewish who came because of idealism and wanted to settle in Israel once the fighting was over. Since they were transport pilots —and other kinds of air and ground crews—they naturally saw themselves creating the basis for an Israeli national airline. The sign in the war was Operation Shahar on 29 September, in which the suscript of El Al was first used, and a document that was prepared by LATA's staff in September about the future of commercial and civilian aviation in Israel. On 15 November Ben Gurion first heard of the troubles in the IDFAF when Chai Issaschar, Aharon Remez and Meir Mardor reported to him about the unrest within the foreign volunteers. It seems that Remez did not realize the seriousness of the situation and mishandled the quarrel, and thus David Ben Gurion had to interfere. The next day the Prime Minister invited Al Schwimmer, and he wrote in his diary:

David Ben Gurion in the pilot seat of a 69 Squadron Flying Fortress.(IAF)

"... Al Schwimmer came. He recruited some 400 aircrews and procured for U.S. about 40 aircraft. In America he told the volunteers based in what he was told by Shlomo Shamir, Teddy Kolleck and their friends (the procurement agents in the U.S.) that they would receive in Israel their expenses and nothing more. In June there was a foreign volunteers meeting. Aharon Remez and Heyman Shamir promised them fair salaries to their specific professions. That promise was not fulfilled. Many suffered from unsuitable accommodations and the like. Ten days ago they had a meeting and decided that if within four days their salary would not be fixed they would all leave Israel. Schwimmer was chosen to deliver the signed requirement to Chel Ha'avir's staff, Swichmmer did not find Remez so he handed over the requirement to Isaschar and Heyman Shamir. Chai answered that there was not any question—the salaries are decided by the government—Schwimmer asked for the government's salaries list and it was handed over to the pilots. The salaries looked fine to the foreign volunteers and the list was accompanied by a list of terms including that each men had to sign that he is serving with the IDF, will obey order, etc.

The men were not happy on two accounts: 1) their standard of living, 2) they do not trust the high level command. They are angry about the list of terms since they've done their best and now they are required to announce that they will fulfill their duty. They choose a committee of three: Harold Livingstonm Ray Kurtz and Norman Monitz. They signed an ultimatum that if the service terms and the salaries will be tied together they will refuse to fly. They did not resist the salaries, but the annexes with the terms. They met Remez, and Schwimmer was there, too. Aharon announced that these are the government terms and that they have to follow them without any changes. Remez was furious about the ultimatum, threatened

Harvard flight line, 1949. (GPO)

with a court martial and ordered them to leave the room. He refused to meet them again without signing on the cancellation of the ultimatum (Schiwimmer and his friends understood their mistake in promoting the ultimatum and in fact Kurtz and Monitz flew Yiga'el to Paris in a C-54 for diplomatic talks in mid-November in spite of the foreign volunteers "strike"—author) and made it clear that anyone who will not condemn the ultimatum should leave the IDFAF. Schwimmer thought that such a humiliation was not necessary. The air council discussed the decision to insist on a formal condemnation to the ultimatum on paper. In a majority of six against five they decided to insist. Livingston did not sign. He is still young. The same day they came to Bristol Hotel to take him to Haifa in order to send him away from Israel. Schwimmer was there and told Livingston not to go since the Chel Ha'avir had the right to relieve him from service but not to send him away from Israel. He (Schwimmer) and his men do not want to be soldiers anymore, they will train the Israelis and leave.

I (Ben Gurion, so far the listener, is now talking) told him that I am discussing the matter with him on an informal basis, person to person. I do not advise him to leave the Chel Ha'avir before the war is over. The support of the American, Canadian and other nationalities in our struggle is not only measured quantitatively, but also in the sympathy of the Jewish people. Their withdrawal before the end of the struggle is a moral defeat for them, the Jews of America, and for us. There is not enough reason for that to happen. Misunderstandings are inevitable when two different worlds meet, the American Jewish youth and the Israeli youth, but they must not be overestimated. The army must have discipline, some commanders are not liked by all soldiers, but all must obey. Otherwise there is no army.

Without exploring deep into details it is obvious that mistakes were made. We must not forward an ultimatum to the army. We may have mishandled the foreign volunteers and we have to improve our handling as much as we can. Our men here highly appreciate the work of the guys from America. I proposed that all the papers: the ultimatum, the condemnation of the ultimatum, the Schwimmer resignation, will all be taken back and forgotten and that we can discuss the problems under the two terms: 1) nobody leaves the Chel Ha'avir before the fighting is over, 2) discipline is essential in any army. Then I will dig into the problems and look for a way to satisfy all.

Al Schwimmer agreed."

And it was as simple as that. Several weeks later a committee was set up that solved the foreign volunteers' problems. They were signed on a one year contract with fair salaries and without formally joining the IDF, and many of them later had key roles in Israeli aviation, such as Al Schwimmer, founder and first chairman of Israeli Aircraft Industries, and Sam Lewis, a El Al'schief pilot for many years

Before fighting broke out once more an administrative step was taken to solve the problem of numbering IDFAF aircraft following the introduction into service of so may new types. The new scheme was quite simple, although it supposed that the IDFAF would never operate more that 100 different types of aircraft and that less than 100 of each type would be procured. Each aircraft was given a four digit number. The first two digits indicated the type, while the remaining two indicated the specific aircraft. The new scheme was introduced towards late November, and as an example, 0419 was the 19th aircraft of the fourth type, which was a Piper PA-11 Cub, while 2303 was the third aircraft of the 23rd type, which was a N.A. P-51D Mustang. Naturally, it took time to adopt that scheme, and many aircraft, especially 101 Sqn. fighters, carried a two digit number that was in fact an unit numbering sequence rather than the IDFAF formal number.

During early December it became evident that Operation Yoav was just not enough. True, the road to the Negev was opened and the Egyptians retreated to the Gaza strip, leaving behind a besieged brigade, but their political leaders still failed to acknowledge the facts and another large scale operation was required to force them to the negotiation table. That was Operation Chorev (Dryness), and indeed it achieved its goals, being the last large scale operation of the War of Independence. However, by early December the balance of aerial forces, at least on paper, concerned the IDF staff. Intelligence reports claimed that the REAF had two squadrons with 42 Spitfires at El Arish and its auxiliary landing strip of El-Riach (its exact location only discovered on 22 December), and two squadrons of newly delivered Macchin MC.205V fighters at the newly constructed Bir Hama air base some 100 km southwest of El Arish. These reports insisted that three Hawker Furies, on loan from the IrAF, were also resident at Sinai. The support elements of the REAF included a heavy bomber squadron equipped with Short Stirlings and two transport squadrons with C-46s and C-47s. The last three units mentioned were all based in the rear at the Nile Delta and were only little involved in the conflict. Although the REAF suffered from low serviceability rates, and the Israeli intelligence numbers may have been a little exaggerated, it was still a respected opponent. Against the REAF, the SAF and IrAF were ignored at the time because of the static situation in the respective fronts. The IDFAF could assemble a respectable force compared to its May 1948 ORBAT, but still much was left to be desired. Following the SS Enterprise shipment from the U.S. of two Mustangs and two Piper Cubs, other shipments arrived. For example, on 21 October five T-6s were unloaded at Haifa, and the ORBAT continued to grow steadily. By mid-December 101 Squadron at Hatzor had five Spitfires, two Mustangs, and six S-199s. At Ramat-David 69 Sqn. still operated the three B-17s alongside 103 Squadron, which had five Dakotas and one operational Beaufighter. At Tel-Nof 106 Squadron operated nine C-46s and two C-54s, while 35 Flight converted from the Norsemans to the newly arrived Harvards that were quipped with a pair of forward firing 12.7mm machine guns and eight 50-kg bombs under their wings for use as dive bombers. The tactic used was to enter a 60/70 degree dive at 9,000 ft and 100 knots, release a pair of 50-kg bombs at 3,000 ft and 300 knots, and use the remaining 3,000 ft to recover from the dive with up to a 10-g pull up! the escape from the target area was made at low level with both machine guns firing to keep heads down. Once at a safe distance from the target the Harvards slowly climbed back to 9,000 ft and started the whole procedure once more. The accuracy of that bombing method was wholly dependent on the pilot's skill since the Harvards were not equipped with aiming devices of any kind. The sectorial squadrons received new equipment, and the Negev and Galilee Sqns. standardized on the Piper Cub, although the faithful Auster AOPs were still in widespread use, while the Tel-Aviv Sqn. still operated Dragon Rapides, Bonanzas, Nords and the like.

Thus, although the order of battle of the IDFAF improved considerably, the numerical ration in fighters was still one over five in the REAF's favor. Since a new offensive was planned a reinforcement was considered of the utmost urgency, and the only available source was the Spitfires in Czechoslovakia. By 22 October 12 more Spitfires were ready for a flight to Israel, but the Yugoslavs refused to grant permission for a refueling stop and the Bulgarians still demanded U.S.$10,000 for the passage of each aircraft. To overcome this difficulty the imaginative Sam Pomerance managed to squeeze more fuel into the Spitfires to allow a non-stop flight to Israel, and by 15 November seven of the converted Spitfires were ready, along with eight others that were equipped for a single refueling stop flight. The Czechs decided to approve only a route that included a refueling stop, but as this stage was not found all the scheme was at stake. Moreover, the Czechs refused more credit and demanded immediate payment. On 29 October the Israeli procurement delegation in Prague reported that up to that date U.S.$7,633.00 had been paid to Czechoslovakia, and that another U.S.$5.5 million should be paid by 15 November. By 25 November 16 Spitfires were ready for flight, but still the green light was not given. In early December the Yugoslavs agreed to allow one more ferry flight, but they too had demands, including the payment of U.S.$300 per ton that was passing through their territory, although that sum was reduced to U.S.$200 after a short negotiation. The Czechs also demanded down payment, and finally the pilots that were to ferry the Spitfires to Israel arrived in Prague on 9 December. However, the preparation took longer than anticipated, and it became clear that the 19 December deadline, the day set for Operation Chorev, could not be met. The IDFAF asked to postpone the operation until the arrival of the Spitfires, but once more the General Staff refused, and it had to do without the awaited reinforcements for the first few days of Chorev.

On 13 December, ten Spitfires packed in crates and hundreds of boxes with spare parts left Kunovice in a train for a Yugoslavian harbor for shipment to Israel, but the ferry flight to Israel of a dozen other Spitfires, named Operation Velveta 2, had to be delayed due to a severe snowstorm. A couple of days later, on 16 December, a C-54 arrived in Prague to collect the Spitfires' equipment and weapons, but its departure was delayed due to a technical problem. The second 106 Squadron Skymaster was waiting for the Spitfires in Yugoslavia to lead them on their long overwater flight. On 18 December Sam Pomerance decided to commence the first stage of Operation Velveta 2 by leading the first six Spitfires on the first leg from Kunovice to Yugoslavia. At 0900 GMT the six Spitfires—three were equipped for the non-stop flight and had no radio sets at all—left Kunovice. The leader was Sam Pomerance, and among the other pilots in the formation were Bill Pomerantz, Sandy Jacobs, George Lichter and Moderchai "Moti" Hod. The last mentioned was an Israeli that received his flying training in Czechoslovakia and had less than 150 hours to his credit, of which less than ten were in Spitfires!. However, Moti Hod did not disappoint his instructor, George Lichter, and eventually became the architect of IDFAF's victory in the Six-Day War. Hod left the Air Force during early 1973 after 25 years of service, including about seven as its commander. The six Spitfires flew in close formation at 10,000 ft through a snowstorm. The first to lose the formation was Bill Pomerantz, almost immediately after take-off, and the second was Sam Pomerance, who lost the formation after one and a half. hours. George Lichter decided to abort the dangerous ferry flight and led the remaining Spitfires back to Kunovice. The first stage of Operation Velveta 2 failed, and the man behind it was killed. Sam Pomerance's crashed Spitfire was found by the Yugoslavs, while Bill Pomerantz made an emergency landing in Yugoslavian territory in which his Spitfire was completely burned out, but his injuries were only light. George Lichter, an American volunteer who served in Czechoslovakia for several months as the instructor of a small group of Israelis that received basic and advanced training there, took over the position as leader of Velveta 2, and on 19 December, with the snowstorms over, two sections of six Spitfires each left Kunovice and made a three hour and forty minute, uneventful ferry flight to the refueling base in southern Yugoslavia. Among the pilots was the leader, George Lichter, John Frederick McElroy, Syd Anton, Cesar Dangott, Arnold Roch, Jack Cohen and the two young Israelis, Moti Hod and Danni Shapira. This later pilot would become the IDFAF Chief Test Pilot years later, also occupying the same post in Israel Aircraft Industries, and he first flew the indigenous Nesher and Kfir fighters.

In Yugoslavia the Spitfires were serviced, but two were found to be fully serviceable for the second, and far more dangerous, leg of the route: the almost seven hours overwater flight to Israel. On 23 December a 106 Sqn. C-54—the other one was still grounded at Prague—led four Spitfires to Israel. Unlike Velveta 1, the improvised fuel system worked properly and the flight was uneventful. However, since Operation Chorav was already at full swing and the REAF presence was quite evident, it was decided to avoid risks and Joseph Doyle was sent out in a Mustang to escort the unarmed Spitfires to their destination. While the four Spitfires were an important reinforcement, they were not yet combat ready since the improvised long range fuel system had to be removed and guns and radio sets installed. That activity took several days, and meanwhile the C-54 returned to Yugoslavia and on 26 December led another six Spitfires on six hours and forty five minutes of uneventful ferry flight to Israel. The pair of Spitfires left in Yugoslavia were airlifted to Israel by 106 Sqn. transports. Velveta 2 was over. It was a success, even if we consider the loss of Sam Pomerance, the injuries suffered by Bill Pomerantz, and the total loss of two Spitfires. The operation gave 101 Squadron an invaluable reinforcement during the heat of Operation Chorev, and moreover, the reinforcement of pilots was also valuable, as during the first days of the offensive the unit had a smaller than usual number of pilots. One of the pilots that arrived with Velveta 2 was Canadian ace John Frederick McElroy, who served as an RCAF Sqn./Ldr during WWII and was credited with 15 kills. McElroy played a key role in the clash between the IDFAF and the RAF on the last day of Operation Chorev.

Due to bad weather the three brigade operation was delayed to 22 December, and the IDFAF part in Chorev was defined as continuous CAPs over the combat zone to secure the fighting units from REAF air attacks, bombing of defined targets, and close air support. Thus, the planning for the first two days was carefully laid up, while for the third day onwards it was thought to take into account the developments in the battlefield. Surprisingly enough, the bombing on the late afternoon of 22 December was executed according to plan. A curious offshoot emerged from the fact that 101 Sqn. had to operate its three types of fighters simultaneously. Since the S-199 had a known tendency for take-off/landing accidents, it was decided that they would take off last in order to avoid the risk of blocking the runway for the others, and the same applied in the landing sequence, so the Avias landed last. The REAF made that day the operational debut of its Macchi MC.205Vs, and a formation was intercepted by a pair of 101 Squadron Spitfires. Rudy Augarten claimed a damaged Macchi. On the same day, the El-Riach landing strip was located, while the new REAF Bir Hama air base was not yet known to the IDFAF. On the night of 22/23 December, the Golani Brigade commenced its deception effort towards Gaza, and early the next morning the REAF retaliated with a pre-dawn attack on Tel-Nof air base, possibly with a Short Stirling at 0450 from high altitude, without causing damage.

As a direct consequence of the renewal of fighting it was also decided to try to eliminate the Falluja pocket, and the Artillery Corps and the IDFAF were instructed to that effort. The growing effectiveness of Israeli aviation forced the REAF to launch several attacks on IDFAF air bases. On the evening hours of 24 December, at 2230, eight 250-kg bombs were dropped in the area of kibbutz MIshmar Ha'amek; three children were killed, while five more were injured. That bombload, probably released by a Stirling, was almost certainly addressed to Ramat-David air base. Also, on 24 December the REAF attacked the Tel-Litvinski barracks, east of Tel-Aviv, and the next day it turned its attention to Hatzor air base and at Kfir Sirkin, which was an ex-RAF station that was not used by the IDFAF because of its proximity to the Jordanian front. Another target that received REAF attention on 25 December was Bet Cobrin east of the Gaza strip. Meanwhile, on 24 December the REAF's air base of Bir Hama was finally located by a 101 Sqn. Mustang, and plans were laid out for a future raid on it. On 25 December the main effort was launched, but it seems that only when the IDF succeeded in capturing Nitzama on 27 December did the aerial activity, as that on the ground, really intensy. That day also saw the last operational

Just after the end of the War of Independence, and with enough Spitfires and Mustangs, the Avia S199 was phased out of service. This picture taken on 6 September 1989 shows the last surviving Avia being carried to the Chel Ha'avir Museum at Hatzerim air base for preservation.

Sprit of 1948—a shirtless Ezer Weizman posing with two friends besides one of 101 Sqn S199s.(GPO)

sortie of an Israeli Beaufighter, which, with a pair of Harvards, gave support to ground forces in NItzana. The capture of that place marked the IDF's first incursion into the Sinai Peninsula, and two thrusts were launched well into Sinai: the first was from Nitzana and north to the Rafah Junction, with the aim of disconnecting the Egyptian army in the Gaza strip from the Egyptian mainland, and the second was from Nitzana westwards towards the key road junction of Abu Ageila. It should be remembered that Sinai Peninsula is a desert, and the one who controls the roads controls the peninsula. Thus, the war was brought for the first time to Egyptian soil. Consequently, REAF activity increased, which meant that air engagements in the area became intense, and the 101 Sqn. pilots exploited their proficiency to the maximum. Moreover, by 28 December most of the ten Spitfires that arrived in Velveta 2 were combat ready, and the number of sorties which the unit could mount grew dramatically.

The first full day of activity for the "First Fighter Squadron" was 28 December, and several combats took place. In the morning hours Joseph Doyle and Gordon Levett (an Englishman) escorted the 35 Flight Harvards that were executing their daily routine of dive bombing the besieged Egyptian at Falluja in a pair of 101 Sqn. Spitfires. That day the bombing was more intense than usual since the following night, 28/29 December, the Alexandroni Brigade was to launch Operation Chissol (Elimination) against the Falluja Pocket, although that operation was a failure and the IDF lost 91 men either killed or missing. The escorting Spitfires detected a formation of eight REAF MC.205Vs, and in the short combat which ensued, Doyle was credited with his first kill in IDFAF service, while three more Macchis were claimed as damaged, two to Levett and one to Doyle. On the same morning at 1045 hrs, a pair of Egyptian MC.205Vs that bombed and strafed an Israeli column near Nitzana were intercepted by a pair of Spitfires on CAP led by Rudy Augarten. Both Macchis were claimed as damaged, and one was seen to leave the combat area in a shallow dive while the other trailed black smoke. A third air engagement in the Nitzana vicinity that morning saw an aerial victory for the REAF. A pair of Auster AOPs from the Negev Sqn. were supporting the IDF troops when both were bounced by REAF Spitfires. Nachom Biran managed to escape, but Tazi Zible, the holder of the IDF highest medal for gallantry, was not so fortunate, and his Auster was downed by the Egyptian fighters. He was killed, as was his observer, Itschak Zamir of the Negev Brigade, who was on his 24th operational sortie. The growing activity of the REAF worried the Southern Front HQ, and two commando raids were hurriedly planned. On 29 December, El Arish was raided with the forces departing from Abu Ageila and capturing the nearby auxiliary landing ground where a single Spitfire was captured intact. On the same day the 9th Regiment left Abue Ageila to the west and raided Bir Hama air base, where five intact aircraft were found, but the request to send pilots to collect them was ignored, and the force retreated east afterwards. These two raids had an effect on the REAF's tactical elements deployed in the Sinai, and on that day Stirlings bombed from high altitude the IDF troops at Abue Ageila without causing any damage. The next day, 30 December, 101 Sqn. Spitfires scored again. The Spitfires were flown by Canadians Joseph Doyle and John McElroy, who jumped a pair of Egyptian MC.205Vs which were strafing IDF troops and which, for a few moments, forgot to check their six. This proved to be fatal, and each Canadian claimed one.

Within hours of the end of 1948, the United Kingdom presented an ultimatum to Israel to retreat from Sinai to the International Border. Not wanting any incident with the British, Ben Gurion ordered the IDF to retreat by dawn on 2 January 1949. However, Yigael Allon (Southern Forces CO), having been so close to defeating the Egyptian army, did not give up easily and received permission to try to blockade the Egyptian forces in the Gaza strip in spite of the disposition of his forces. That move did not materialize completely as the Egyptians reacted strongly, but it delayed the end of Chorev for several days, and it was Allon's insistence that was to culminate with the Egyptians agreeing to negotiate with the Israelis for the first time. On the evening of 31 December Allon boarded Piper Cub in Ber Sheva, piloted by Abraham Portugali, in order to fly to Tel-Aviv and discuss the withdrawal of his force. The small airplane arrived in the Tel-Aviv area just when a small force of Egyptian warships shelled the area with about 20 rounds. For obvious reasons the lights at Sde-Dov were not turned on, and only after several requests they did light, briefly, and the Cub landed. Several thousand kilometers to the northwest, the grounded C-54 at Prague, already serviceable, left on 1 January, and despite the fact that one of its engines did not work and had to be shut down, the captain, Larry Rabb, decided to continue. Just like in the Halifax incident in late July 1948 the lights at Tel-Nof were not lit up for the fear of an REAF raid, and, like John Harvey, Larry Rabb diverted to Tel-Aviv, where the lights were also turned off, and just before dawn, 2 January, Robb belly landed the transport on the beach with only two working engines. The Skymaster was destroyed, but the crew escaped unscathed and the cargo (spare parts for the Spitfires) was saved.

The anticipated end to Operation Chorev intensified IDFAF operations even more, and on 31 December the Falluja Pocket was bombed three times. A 101 Sqn. formation of three Spitfires raided Bir Hama air base at 0745 hrs, with two strafing and the third pro-

Two deHavilland DH.89A Dragon Rapides of the Tel-Aviv Squadron, escorting a convoy bound for Jerusalem during the War of Independence. (IAF Magazine)

viding cover. They claimed two MC.205Vs destroyed on the ground, including one which was taxing, while the Canadian Danny Wilson was credited with a confirmed kill on an airborne Macchi. In the next night, 31 December/1 January, the Falluja Pocket was bombed four times—Gaza three times, and Dir el Balah, south of Gaza, once. On 1 January Spitfires were tasked to dive bomb a railroad junction in the Rafah area, and the next day two Spitfires attacked a moving train in northern Sinai, but two of the bombs they dropped did not explode. The Spitfire which had been captured in the emergency landing ground near El Arish was towed by a tractor all the way back to Israel. In the evening hours of 2 January the REAF made its last attack against civilian targets in Israel when a Stirling bombed Jerusalem from high altitude; four citizens were injured. On 4 January the Hariel Brigade commenced its attack on the Rafah junction, and the air activity intensified once more, with the attention turned to that vital area where fierce fighting took place. Although the brigade effort failed, the main road from the Gaza strip to mainland Egypt was cut by the same brigade just south of Rafah junction. At 1620 hrs two B-17s bombed Egyptian barracks and depots in Rafah; the pair was to be escorted by two Spitfires, but the fighters could not locate the bombers, although their pilots observed the gray smoke associated with the bombing. Curiously enough, a pair of Spitfires on CAP in the area made visual contact with the Flying Fortresses.

Earlier that day at 0950 a single B-17 bombed Egyptian concentrations near El Arish with ten 250-kg bombs. Three Harvards dive bombed the Egyptian positions at Rafah at 1130 in support of the attacking brigade. As usual, 106 Sqn. makeshift bombers commenced operations after darkness fell, and at 2350 a single Commando bombed Gaza with 24 50-kg bombs. The next day 101 Sqn. fighters attacked a train just south of Khan Yunis and damaged the engine, while a mixed formation of a Spitfire and a Mustang from the same unit intercepted at about 1200 three REAF MC.205Vs that were at 7,000 ft over Rafah and about to dive bomb the IDF troops in the area. The Israeli fighters jettisoned their ordnance, and one of the Macchis fled immediately. Of the other two, one was claimed by the Mustang pilot, the American Cy Feldman, as a confirmed kill. On the debit side, an Auster AOP was lost during the day when it crashed on landing at Ber Sheva, but the pilot was unhurt. In the evening hours of 5 January a single REAF Stirling bombed Lod airport, which was not used by the air force, but was occupied by the 82nd Regiment from the IDF. One of the bombs hit the mess hall, and one soldier was killed and two others wounded.

In the late evening of that historic day, 5 January 1949, the message that the Israeli leaders were waiting for several months finally arrived. It was sent on 4 January, the very same day that the Hariel Brigade blocked the main road from Gaza to Sinai and threatened to besiege the Egyptian army in the Gaza strip. Sent from Cairo to the U.N. at New York, the message was short and clear: if fire would cease by 5 January at 1400 hrs GMT, the Egyptian government would instruct its representatives to immediately open negotiations with Israel under the U.N. sponsorship. A sand storm on 6 January stopped all operations, and finally, under U.N. pressure and with the approval of both sides, it was decided to cease all fire on 7 January at 1400 hrs GMT (1600 local time). The War of Independence was not yet over, since on the morning of 7 January the IDFAF was instructed to exploit to the full the last hours before the formal cease-fire, and it was that decision that paved the way for the first aerial clash between Israel and a world power, and like the outcome of the clash against the Russians twenty-one years later, the Israelis gained the upper hand.

CHEL HA'AVIR VERSUS ROYAL AIR FORCE

The late December 1948 British ultimatum and threat to enact the 1936 mutual treaty with Egypt was based on intelligence gathered by a series of reconnaissance flights. The British commanders in the Middle East ignored the Noon Charlie fate, or it may be that its loss was not attributed to IDFAF action, and reconnaissance flights were dispatched to northeast Sinai to gather first hand information concerning the Israeli advance. It is true that the RAF aircraft had every right to overfly the area, which was Egyptian territory, but it was a battlefield, and as such they had to be ready for the worst. On the morning of 7 January two RAF formations left the British air bases in the Suez Canal Zone for another "routine" recce sortie. For one of the formations, a single Mosquito PR.13 of No 13 Sqn. from Kabrit escorted by four Tempest F.6s of No 6 Sqn. based at RAF Deversoir, it was the usual uneventful flight. For the other formation, four Spitfire FR.18s of No 208 Sqn. from RAF Fayid, it was to be a one-way mission, as none of them returned home. The foursome Spitfire flight left Fayid at 1100 hrs led by F/O G.S. Cooper. The other pilots were P/O McElhaw and F/Sgts Close and Sayers. They were briefed by No 208 Sqn. C.O. Sqn./Ldr Morgan, and their orders were clear; they were to gather information about the position and location of Israeli forces in northeast Sinai by surveying the Nitzana-Rafah area. Their route was to Abu Ageila, where the four had to split with P/O Cooper and his wingman F/Sgt Close working as the lower element and P/O McElhaw and his wingman F/Sgt Sayer as the high element to provide top cover. From Abu Ageila the route turned east to Nitzana and then north, along the international border, to Rafah, where the formation had to turn west and return to its base. The Spitfires were armed with ammunition for their guns, and the leaders of both elements were credited with confirmed kills some seven months. F/O Cooper and McElhaw were among the No 208 Sqn. CAP which downed four of the five REAF Spitfire IXs that attacked their Ramat-David base on 22 May 1948. Thus, they were confident enough and probably trusted highly their Griffon-engined Spitfires, which were superior on all aspects to the REAF and IDFAF Merlin-engined Spitfires. However, one parameter was missing in that equation—the human factor. While RAF fighters were flying east two separate events took place which were to influence the fate of the four British pilots. REAF MC.205Vs were quite active that morning. At about 0930 hrs a 101 Sqn. formation escorting Harvards that were to dive bomb Dir El Balah engaged eight Macchis, and Joseph Doyle, who flew a Mustang, claimed one destroyed and another one damaged. Later on a formation of Macchis attacked IDF troops in the Rafah junction area, just fifteen minutes before the RAF formation arrived at the scene. The attack had two outcomes—it made the Israeli forces trigger happy, and as vehicles were damaged, three black smoke columns marked the location for miles.

The second event took place on Hatzor air base at 1135 hrs when a 101 Sqn. CAP took off to cover the IDF troops at Rafah. The formation was to include three fighters, two Spitfires and a Mustang. However, the P-51D went unserviceable and its pilot, Lee Sinclair, who had been an RAF Wing Commander, stayed behind. The two pilots that did leave Hatzor's concrete runway behind were the Canadian John Frederick McElroy and the American Slick Goodlin, and they turned southwest at full throttle, leaving behind a thin trail of exhaust smoke from the mixture-rich Merlins. By pure coincidence the three parties, the RAF, the IDFAF formation and the nervous ground forces at Rafah, met exactly at the same time and at the same spot. The RAF pilots were attracted by the rising columns of smoke, and were welcomed by small arms fire. At exactly the same moment, McElroy and Goodlin, also attracted to the area by the smoke, observed the British Spitfires. They knew they were the only Israeli fighters in the area and that any formation larger that two or three aircraft should be enemy. Moreover, the possibility of a British intervention was not forecasted by the IDFAF and the pilots were not instructed to explore the identity of the adversary, but only to insure that it was destroyed. The No 208 Sqn. top cover element fulfilled its purpose by warning the lower element about the presence of bogies at "3 o'clock", but that warning came too late since almost simultaneously the ground forces and the 101 Sqn. fighters inflicted their damage. The IDF ground fire accounted for F/Sgt Close's Spitfire, and F/A Cooper was able to see his wingman bailing out of his doomed fighter before McElroy's first bursts hit him. Cooper was wounded in a leg, but he tried to turn with McElroy and even fired a pair of bursts before he also had to abandon his aircraft, and McElroy saw him (once on the ground) running west towards Egyptian held territory. While McElroy was dealing with Cooper, the upper element of the RAF formation climbed into the sun closely followed by Goodlin. The superior performance of the FR.18s worked to the RAF pair's favor, and Goodlin was left behind. At 7,000 ft F/O McElhaw broke, dived at the Israeli fighter, and opened fire. Goodlin was experienced enough to observe the tracers. He then maneuvered his opponent into a dogfight in which the superior performance of the FR.18 could not be exploited to the full. In such a kind of turning combat the better pilot survives. At that stage Goodlin clearly observed the RAf rounds, but, flying the slower Spitfire LF.9 he could not break contact, and gained the upper hand in the turning contest, and with a few short and well-aimed bursts, McElhaw parachuted from his mortally wounded Spitfire, thus joining the unconscious F/Sgt Close in the hands of the Israeli soldiers that watched closely the air combat above them. And then there was Sgt Sayers. After

One of the early Spitfires of 101 Sqn revving up at Hertzlia airstrip prior to taking off for a recce sortie late in the summer of 1948.(IAF Magazine)

Yugoslavia, December 1948 during operation Velveta 2, *with two Israeli pilots of 101 Sqn, Dannin Shapira sitting on the nose and Moti Hod below. (IAF)*

finishing the leader of the RAF formation, McElroy climbed to gain height advantage and to see where his wingman was. He noticed that Goodlin was doing quite well and observed the last No 208 Sqn. Spitfire down below. McElroy quickly dived at it and opened fire. Poor F/Sgt Sayers probably never knew what, or who, hit him, and his Spitfire crashed into the desert floor.

At 1227 hrs the pair of Israeli fighters landed at Hatzor and reported that they had downed three RAF Spitfires. No one believed them—it was too fantastic to be true. But, some fifteen minutes later the Israeli forces at the Rafah area reported back that three British fighters were downed, one by small arms fire and two by IDFAF fighters (it seems that the ground forces missed the last kill since Sayer's Spitfire crashed in Egyptian held territory). Moreover, they reported that they had captured two of the British pilots; McElhaw, and Close, who was given medical care after losing consciousness because he bailed out at very low level and his parachute opened shortly before hitting the ground, which he did very hard. During the initial interrogation McElhaw claimed that he was shot down by another aircraft, and after his Spitfire caught fire he had to abandon it. Close reported that while flying at low level his Spitfire was struck by small arms fire and after it caught fire he had to jump. The news were not welcomed with enthusiasm since no one wanted to get messed with the British and especially at that stage when the long awaited cease-fire was so close. However, the IDFAF could not leave the ground forces in the Rafah area to the mercies of enemy attacks, and the order was passed down to 101 Sqn. at Hatzor to keep on CAPs, but beware of British fighters. That order left too much open space to maneuver for adventurous fighter pilots smelling the chance of achieving "trade", and 101 Sqn. started to send CAPs to the area one after another. At 1320 a pair of Spitfires were dispatched from Hatzor on CAP, and at 1410 a pair of Mustangs, led by Lee Sinclair, took off for the same mission. However, they did not find anything, and it was the fate of the 1530 hrs CAP to meet RAF fighters that were dispatched to search for the missing No 208 Sqn. formation. Back at Suez Canal stations the British were truly worried by the fate of the missing Spitfires. The Mosquito and its Tempest escort returned, but the Spitfires did not and it was decided to send yet another formation to obtain first-hand information concerning the fate of the Spitfires. To avoid risks that formation consisted of four Spitfire FR.18s and fifteen Tempest F.6s. The fighters left their bases in the Suez Canal Zone at 1500 hrs and proceeded east towards the land battle zone where a combat formation was adopted, again to minimize risks. This consisted of three elements: the four Spitfire FR.18s made up the lower element; the medium element consisted of eight Tempest F.6s from No 213 Sqn. led by Gp/Capt Anderson; and the high element was made up of the remaining seven Tempests from No 6 Sqn. led by Sqn./Ldr Crowley-Milling, DSO, DFC and bar. At 1530 a formation of four Spitfires, quite large for 101 Sqn. standards of the period, left Hatzor for the last CAP of the day, and in fact the last of the war against Egypt in northeast Sinai. The cease-fire was already in effect, and the formation orders were to act only as a show of strength. Ezer Weizman led his wingman Sandy Jacobs, and the second pair was composed by Cesar Dangott and his wingman, Bill Schroeder. At 7,000 ft over the international border, Weizman noticed eight aircraft at "12 o'clock." He banked his wings to attract the attention of his mates and climbed for height. At 8,500 ft the RAF formation, flying in two quartets some 1,500 ft lower, passed underneath the 101 Sqn. formation, probably unaware of its existence. The Israelis turned right and attacked the British formation, which was in fact the top cover, having failed to observe the other elements below. In the first pass Bill Schroeder downed one of the Tempests, probably hitting the pilot because the fighter lost control and spun and was found by Israeli ground forces with its pilot, F/O David Crossley Tattersfield, dead in the cockpit. Following the first pass the usual mess began, with the RAF fighters dropping their auxiliary fuel tanks to add to the confusion. In the two-minute combat that followed Ezer Weizman got in behind a Tempest and fired a long burst at it while the other pilots damaged yet another two Tempests before the British fighters broke contact by virtue of their far superior top speed. The four Spitfires returned to Hatzor, and the only damage they sustained was a few holes in the spinner of Weizman's Spitfire. Moreover, it was claimed that Weizman's victim crash-landed at El Arish, and although its pilot escaped unhurt, the Tempest was completely wrecked. The 101 Sqn. pilots—most of them either served with the RAF or in other air forces that fought alongside the RAF against the Nazi enemy in WWII—were not happy with the incident, although they fulfilled their duty to provide cover for the ground forces. Contrary to what was feared, there was no British retaliation, and instead the Foreign Office forwarded the Israeli government a demand for an economical compensation for the loss of personnel and equipment.

By then Operation Chorev was over and the war was nearing its end. During Chorev the IDFAF had finally matured. It executed the pre-planned bombings, achieved total air superiority over the battlefield, and supported their own forces. During the two and half weeks of this operation, only one aircraft was lost to enemy action, Zibel's Auster on 28 December, while claiming six REAF Macchi MC.205Vs, three RAF Spitfire FR.18s and one Tempest F.6 destroyed in air combat. In addition, two Macchis were destroyed on the ground at Bir Hama, while at least seven were claimed as damaged in air combat. Ground forces accounted for another RAF Spitfire, and a single Egyptian Spitfire was captured at El Arish, although it was damaged when it was towed back to Israel. The total number of sorties flown by the IDFAF was 243.

Following Chorev the limelight turned to the Greek island of Rhodes, where the Israeli and Egyptian delegations, under U.N. supervision, met for cease-fire talks. On 24 February 1949 the cease-fire agreement between the two countries was signed, and although it was not a peace treaty, it was a major breaktrough closely followed by similar agreements with Lebanon (23 March), Jordan (3 April), and Syria (20 July). The Iraqis, who did not share a border with Israel, withdrew their forces, but did not sign any cease-fire agreement. Amidst the political effort to at least temporarily end the conflict, the last operation of the war took place to ensure that the state of Israel should have free access to the Red Sea in accordance with U.N. Resolution 181 from 29 November 1947.

The operation was aptly named Ovda (Fact). Early in March, two brigades were to force their way from Ber Sheva in the central Negev to Om Rashrash (now Eilat) on the beautiful beaches of the Aqaba Gulf, while a third brigade was to advance from Sdom along the western shore of the Dead Sea. Evidently, the IDFAF provided transport support through 103 and 106 Sqns., as well as fighter cover, and observation by the Negev Sqn. Piper Cubs, while the B-17s of 69 Sqn. stood at readiness for search and rescue operations. An advanced landing ground was specially prepared some 50 km north of Om Rashash and code named Abraham Field. During the whole operation, which culminated in success, only one aircraft was lost due to a take-off accident from Abraham field—a C-46 Commando (4X-ACG) whose captain was Pessha Tal—but nobody was hurt. The Jordanians accepted the outcome of Operation Ovda and signed the cease-fire on 3 April. The conflict, known in Israel as the War of Independence, was over, but peace was not at hand, and an attitude existed that forced the Middle East into an arms race which has continued into the present.

A pair of 101 Squadron Spitfires escorting a B-17G mission as seen by the right side waist gunner. (IAF)

Chapter 7:
Post War and Consolidation

By the end of the war the IDFAF was arguably the best air arm in the Middle East. It had started as the Sherut Avir of November 1947, with its small number of obsolete light-planes and a few dozen enlisted men supporting the Haggana forces until the British withdrawal and the creation of the state of Israel in May 1948. Then the appearance of the Arab air arms forced the newly-named Chel Ha'avir to turn its activities to the night, at least until the arrival of the first fighters, the Avia S-199s, and even then it could only achieve local air superiority over the densely populated Tel-Aviv area. Total air superiority was only achieved towards the end of the conflict with the Spitfire LF.9s and P-51D Mustangs, as well as with the help of top quality foreign volunteers. It was a team of colorful experts that on one hand achieved the much desired air superiority and on the other hand taught the young Israelis of 101 Sqn. how to become successful fighter pilots, a message that was passed within the IDFAF and has created some of the most successful aces of the jet age.

By the end of the war summaries could be made; during the roughly 18 months of fighting thc air force had ups and downs, successes and failures. It lost 24 of its pilots and 24 other air and ground crews killed. In addition, two pilots were listed MIA, and three pilots and one observer/photographer were taken prisoners. In total, the IDAF lost during that time period 69 aircraft to all causes in the theater of operations and abroad. On the other hand, the Chel Ha'avir and the IDF ground forces claimed a total of 34 adversary aircraft destroyed in air combat, air raids or AA fire, including five RAF fighters. The top 101 Sqn. scorer was Rudy Augarten with four and half victories—the half was an REAF C-47 downed on 11 November 1948 shared with Boris Senior. Three other pilots got three kills each: Modi Allon, Joseph Doyle and John McElroy. The total number of air-to-air victories for 101 Sqn. was 22, so these four pilots claimed more than 50% of the total score!

During the conflict the air force grew from a single squadron of light-planes to a complex organization with well over 6,000 enlisted men and women, however, of that total only about 150 were pilots. Furthermore, most of the achievements had been obtained by the foreign volunteers, and on 2 February 1949 there were 291 in IDFAF service, which amounted to less than five percent of the total strength, but they played key roles. So, it was only natural that once the war was over the highest priority was given to train Israeli pilots so they could take over these places. By the early summer of 1949, the IDFAF went through it first major reorganization, and many foreign volunteers departed back to their homelands and painful budget cuts were made to adjust to the economic situation and

The Boeing Stearman Kaydet was the standard primary trainer during the late forties and early fifties.(IAF)

Avro Ansons were used by the Flying School as multi-engine and navigation trainers. (IAF)

the influx of immigrants. To cut expenses some of the air bases were closed. A few units disbanded and others amalgamated; for example, Hatzor was closed—in fact, it was handed to the IDF, and 101 Sqn. under the command of Ezer Weizman, which had replaced Syd Cohen on 1 April 1949, moved to Ramat-David, which was to be the unit's base as Wing 1. The two resident units moved to Tel-Nof, 103 Sqn. with its Dakotas and 69 Sqn. with its B-17s. At Tel-Nof, 106 Sqn. disbanded and handed its C-46s to 103, while 35 Flight also disbanded and its Harvards were handed to the newly created Flying School. Also at Tel-Nof, the Tel-Aviv, Negev and Galilee Sqns. were all amalgamated to create 100 Squadron, which nevertheless maintained a flight at Ber Sheva in the Negev as needs dictated. Finally, the ex-RAF station at Kfar Sirkin, ten miles east of Tel-Aviv, was activated as the home of the Flying School named after Modi Allon.

Thus, in early 1950 the air cadets were sent to the Flying School at Kfar Sirkin that was finally organized and commenced their training there. However, they were not the first to graduate since the groups that had trained in Czechoslovakia and Italy were gathered once more for the advanced stage, and after almost six months of training the first formal IDFAF Flying School graduation ceremony took place on 25 July 1950, with Hugo Marom as the first cadet to graduate with excellency. This was followed by the graduation ceremony of Flying Course number 2 on 14 December 1950. Among the graduates of that second course were: Benjamin Peled, who was IDFAF commander in 1972/1977; Yeoash Tsidon, who left the air force as Aluf Mishne; Omri Arieb, who was killed a few years later in an air collision while flying a Mustang in an unofficial aerobatics display; and Abraham Hacohen, who was killed as IAI's Chief Test Pilot when the Arava prototype crashed in late 1970. The train-

Air Parade during the graduation ceremony of the first class at the IDFAF Flying School at Kfar Sirkin, 25 July 1950. (IAF)

When enough Spitfires became available a second fighter outfit was commissioned, the was 105 "Scorpion" Squadron. During its life this unit flew Spitfires, Mustangs, Super Mystères, F-4E Phantoms and finally F-16C/D Falcons.(IAF)

ing syllabus adopted in the Flying School was based on the RAF scheme and included a primary stage in Kaydets and a basic stage in T-6s. A small number of Avro Ansons and Airspeed Consuls were used for future multi-engine pilots, as well as for navigators and wireless operators. However, the Stearman was not considered adequate, and it was decided to procure a new primary trainer. After an evaluation the Dutch Fokker S.11 was selected and a single example was brought to Israel for evaluation. 41 were ordered during the summer of 1950, but it proved to be a huge failure. It was disliked by the instructors, who were used to the open cockpits of the Stearmans, and it had a weak landing gear. Thus, the S.11 had in Israeli service the sad fate that it was replaced by the same type that it was to succeed!. The Stearmans returned to the primary squadron at the Flying School while the Fokkers were transferred to 100 Sqn., where they were little used until completely withdrawn from use after the Sinai Campaign of 1956, years before the faithful Stearmans were replaced by the Piper Cubs in the Flying School.

A group of Israeli pilots during a party in the early fifities.(IAF)

Dan Tolkowski was the IDFAF fifth commander, and was in charge during the Sinai Campaign of 1956. (IAF)

But procurement of new trainers was not the answer to IDFAF problems at the time. While the Arab air forces were busy acquiring all types of aircraft, the Israelis tried to imitate them in spite of the fact that their basic problem at the time was not the acquisition of more aircraft but lack of skilled air and ground crews and poor administration. In a report to the Minister from 20 June 1949, Aluf Aharon Remez illustrated a gloom picture: of five Dakotas only two were serviceable, and of 14 Stearmans four were serviceable. Other types followed the same pattern: ten out of 17 Piper Cubs, four out of 12 Harvards, and so on. A poor serviceability rate of 40 percent, combined with 4,950 men and women in strength, including some 300 foreign volunteers. The only good news was that 101 Squadron, under the leadership of Rav Seren Ezer Weizman, in a true report for 1 June 1949, showed 27 of the squadron's 29 fighters were serviceable, including 19 out of 20 Spitfires, five out of six S-199s, and all three Mustangs. However, the problem was not fighters, as another 27 Spitfires and two Mustangs were in Israel still in crates, while another 12 Mustangs procured in the U.S. were awaiting shipment to Israel, but fighter pilots. On 14 June 1949 they had 25 fighter pilots in strength, of which seven were experienced foreign volunteers and 18 were Israelis. However, most of the Israelis were not experienced and eight of them were in fact in operational conversion. The problem was even deeper with B-17 pilots: only two were Israelis and another seven were foreign vol-

During the early fifties the B-17Gs which were constantly updated and kept in service, although at the eve of the sinai War, 69 Squadron had disbanded as an economy measure only to be reformed for that conflict.(IAF)

unteers; of transport pilots, only 12 were Israelis and another 25 were foreign volunteers; and of the navigators, only eight were Israelis and another 14 foreign volunteers, etc. Thus, the IDFAF had to act at the time on two levels, training and procurement. However, it seems that only after Aharon Remez left his post, the procurement effort really took off the ground. It seems that Remez ignored the daily life and concentrated on rather imaginary schemes, a fact that resulted in the bad administration within the air force, the poor serviceability rates and the fact that the IDF CoS warned the Minister that the internal situation within the Chel Ha'avir was deteriorating. Nevertheless, all that did not deter Remez from presenting imaginary plans to Ben Gurion at the 20 June meeting, including a scheme of no less than seven wings at the time when the IDF strength was being cut drastically. This plan comprised a total of 292 aircraft, although this scheme was not put into real practice until some ten years later or so when Ezer Weizman became the IDFAF commander. Remez understood that he had to clear the post, otherwise the air force would suffer much more, so in late 1950 and at the early age of 31, after serving for about two years as its commander, Aluf Aharon Remez presented his resignation. Remez chose to leave on the same day that the second flying course was to graduate, which was the first in which the full syllabus was applied (primary/basic/advanced). Also, the fact that the graduates chosen to be fighter pilots should attend a Operational Conversion Course was not ignored, and during the summer of 1950, the second fighter squadron, numbered 105 and nicknamed "Tayeset Ha'akrav (the Scorpion Squadron)—after its crews which featured a scorpion—was finally activated with Spitfires alongside 101 Squadron at Ramat-David. 105 Sqn. Spitfires were distinguished from those of 101 (which was known as "Tayeset Ha'rishna"—the First Fighter Squadron) by virtue of their yellow-black stripped rudders and yellow spinners against 101's red spinners and red-white stripped rudders. The new squadron was to serve as the IDFAF OTU until 1956, and by the time the second flying class graduated on 14 December 1950, it was ready to commence an operational conversion course while still maintaining its first line operational commitments. 105 Sqn. first C.O. was Rav Seren Richard "Grisha" Brown.

To succeed Aharon Remez as IDFAF commander, 38 years-old Aluf Shlomo Shamir was chosen, who, although having a PPL license, did not serve a single day in the air arm. The Chel Ha'avir top officers resented Shamir's appointment and all resigned, and he accepted all except that of Aluf Mishne Dan Tolkowski. Shamir had to retire after only eight months in his post due to his poor health. That short period of command was commenced by two important steps. First, he changed the ranks, which up to his arrival were similar to those of the RAF, to those used by the IDF. For the sake of clarity this narrative has ignored so far that short-lived system, but just to illustrate it Rosh Tayeset (Squadron Leader) was equivalent, and changed to Rav Seren; Rosh Kanaf (Wing Commander) was changed to Sgan Aluf; and Rosh Lahak (Group Captain) was to become Aluf Mishne. This system continued unchanged until a new rank was introduced after the Six Days War of June 1967—Tat Aluf (Brigadier General). The second step consisted of moving the IDFAF staff from its location at Jaffe, on the southern suburbs of Tel-Aviv, to the muddy tent camp at Ramleh, east of Tel-Aviv. Accompanying the staff was the 100 Sqn. that left Tel-Nof posted to Ramleh, thus providing much needed liaison services with its Piper Cubs.

It was during Shamir's short period of command that the IDFAF saw its first operational sorties since the end of the War of Independence. All along Israeli borders local skirmishes were almost a daily routine, usually with the Arabs attacking and the Israelis retaliating. During one of those skirmishes on the morning of 5 April 1951, an Israeli patrol, which was at Hammat Gadder, the point at which the Israeli/Syrian/Jordanian borders met, was attacked by Syrian

Egyptian Air Force Yak-11 advanced trainer whose pilot defected to Israel. He was given a new identity and settled in Argentina, but was found there by Egyptian secret service men and murdered.(IAF)

deHavilland Mosquitos served in considerable numbers with the Chel Ha'avir, both in fighter-bomber, night-fighter, recce and training duties, mostly serving with 110 Squadron. (IAF)

forces and seven Israeli frontier guards were killed. The decision was taken to retaliate the same day with an air attack. Actually, retaliation raids by the IDFAF are the norm now, but then it was a revolutionary step, and although that first attack was not successful, it set the shape of things to come. Nine aircraft took part in the operation, although only eight were armed, while the ninth, a Harvard flown by Seren Benjamin Peled and accompanied by Sgan Aluf Paul Kaider, was to serve as a communications relay. The other aircraft operated in two formations of four each. Rav Seren Meir Rof, 101 Sqn. C.O., led four Mustangs armed with rockets, while Rav Seren Menachem Bar, 105 Sqn. C.O., led four Spitfires armed with bombs. Both pilots had recently appointed to command their respective squadrons. The attack was carried out during the late afternoon of the same day when the setting sun was behind the aircraft, which made their runs from west to east, directly into the eyes of the Syrian forces. The four Spits were the first to attack, and the target was the local police station at Hammat Gader, which was dive bombed with poor results. Anyway, it is very unlikely that the relative light bombs with which they were armed could inflict any kind of significant damage to such a concrete building even if direct hits were scored. After the Spitfires came the four Mustangs, which fired their rockets at a bunker and a group of tents, with two rockets claimed to have struck a bunker while a fire reportedly broke out in the tent area. The Syrians replied with small arms fire, but none of the attackers were hit. The damage caused by this pioneering attack was light, but the responses it provoked in the world were so intense that it was decided, for political considerations, not to execute any more retaliation air attacks. Thus, it was quite surprising when a month later the IDFAF participated once more in a border skirmish with Syria. On 6 May a 105 Sqn. formation led by none other than the legendary Rudy Augarten—who had returned to Israel to command Ramat-David air base and was known by his Hebrew name, Zev Carmi—while the other three pilots were Honda Costa, Zev Liron and Jacob Shalmo, all very young aviators; the appearance of the Spitfires saved the day for the hard-pressed Golani Brigade.

Thus, Aluf Shamir experienced a short but intense command period, but he was not able to pursue his ambitious plan to trim IDFAF's oversized staff and that distinction fell to the next (fourth) Chel Ha'avir commander, Aluf Chaim Laskov. As its own officers were not mature enough, this officer, an expert in the field of armored warfare, was asked in August 1951 to take command of the air arm. Laskov was born in 1919 and lost his father at the age of 11 when he was murdered by the Arbas in Haifa. His career was quite similar to that of Shamir—he enlisted in the Haggana, and then in the British Army during WWII, where he reached the rank of Major, and finally in the IDF. After his period in the Chel Ha'avir Laskov continued his army career to become the commander of the IDF Armored Corps during the Suez Campaign of late 1956, and in January 1958 he succeded Raf Aluf Moshe Dayan as the IDF Chief of Staff, retiring three years later in January 1961.

Laskov was a character. He could not bear bureaucracy and waste and he immediately set out to learn IDFAF's specific needs,

A P-51D Mustang with 101 Squadron markings, its natural metallic finish was replaced by a camouflage scheme when the first Gloster Meteors arrived. (IAF)

a process that culminated in a report which was not flattering to the air arm at all; its staff posed the problem from which most of the others emerged: lack of leadership and too much, far too much bureaucracy. Laskov learned that the staff numbered 1,253 men and women. It was a giant (something like a quarter of IDFAF personnel strength at the time), and it was swallowing budgets and producing unnecessary bureaucracy. As a result, morale was low, the discipline standard was low, combat readiness was unacceptable, and serviceability rates were low. The poor administration led to logistics problems, the fuel stock allowed only fifteen days of full fighting, while the stocks of rockets and incendiary bombs were dangerously low and spare parts were scarce. Laskov summed up that the air arm could probably fight for only two consecutive days if an all-out conflict would flare up. But even in those two days he doubted about the ability of the IDFAF Staff at Ramleh to control the operations of the first line units because of the poor state of the communications network. Although Laskov was not a pilot, he was able to conclude that many pilots were still not combat ready, although they flew their aircraft well during peacetime. He commenced his painful surgery and when he left the air arm two years later the staff numbered only 525 men and women, a remarkable achievement by all standards.

A Mosquito FB.6 while in service with 110 Squadron.

This was in short the youth pains of the IDFAF, an air arm which was born in war, and which came to the surface after the conflict ended. Now the way taken was the correct one, but shortly after that war an arms race developed in the Middle East that would have a profound effect on Israel and its air force, as the new country security was again in jeopardy. Egypt, and to a lesser extent the other Arab countries, commenced a worldwide procurement effort that was to culminate in the "Czech Deal" of late 1955 and the defeat of Egyptian forces in Sinai during late 1956.

The traditional arms supplier to Egypt was the U.K., but because of the U.N. embargo on the Middle East the British, although fulfilling past contracts, could not respect new ones, and Egypt turned to Italy for the supply of fighters. However, the German-engined Italian fighters, the Macchi MC.205V and Fiat G.55, were somewhat ineffective and were quickly withdrawn from use just a few years after the conflict was over in preference for the new equipment supplied by Great Britain once the U.N. embargo was lifted on 11 August 1949 after the cease-fire agreements were signed at Rhodes. Israeli intelligence reports of the period were quite hysterical, but one thing was certain; as early as August 1948 Egypt ordered its first jet fighters to become the first Middle East country to operate them. Moreover, because the U.N. embargo was still in force, that initial order, for just a single seat Meteor T.7 and a pair of Meteor F.4s, was fulfilled only towards the end of 1949, being followed by another ten F.4s, five T.7s, 12 F.8s and six NF.13s. However, the delivery period of these aircraft spanned well over six years, until 1955, and was parallel to the delivery of 62 Vampires, including 49 FB.52s, 12 T.55s and a single FB.5, making a total of 98 British-supplied jet fighters. Concerning piston aircraft,

The Mosquito PR.16 was the main recce platform in the Israeli Air Force, complemented by the later jet powered Meteor FR.9, until the arrival of the Vatour IIBR from France. A total of eight of these were acquired between 1951 and 1956. (IAF)

Almost one hundred Mosquitos, mostly FB.6s, T.3, T.33, PR.16 and NF.30 were purchased by the Chel Ha' avir, serving with at least four squadrons until replaced in the late fifties. These were 110, 119 and an Operational Training Unit, 109 Squadron. Illustrated is a NF.30. (IAF)

A 119 Squadron pilot poses beside a Mosquito NF.30. Note the characteristic bat insignia of this squadron on the back of his flying suit. This unit was later re-equipped with Meteor NF.13s, although it retained a few Mosquitos for training for a few more years.(IAF)

Egypt received 20 Spitfire F.22s and a dozen Sea Furies, as well as nine each of the Halifax and Lancaster heavy bombers. Meanwhile, the Syrians had ordered a dozen Fiat G.55s from Italy and 23 Meteors in the UK. Thus, Israel was facing re-equipped adversaries without the ability to procure a suitable answer. During 1950 the spearhead of the IDFAF consisted of its two fighter units at Ramat-David. The 101 Sqn., mainly equipped with Mustangs (a total of some 30 had been shipped from the U.S. since late 1948) also operated the faithful Spitfires, while 105 Sqn. was fully equipped with Spitfires, and a single bombing unit, 69 Sqn., had only three B-17s. The missing link since the withdrawal of the last Beaufighter during 1949 was a twin-engined attack bomber. The gap was filled by the procurement of large quantities of DeHavilland Mosquitos from France. The Armée del'Air had operated a large fleet of the "wooden wonder", and by 1950 most of them had been withdrawn to the storage and scrapping center at Chateaudun. The Mosquitos were deprived of all valuable equipment—radio sets, weapons and radar—and awaited their ultimate fate. However, a small number were operational in France and northern Africa. Aluf Mishne Benjamin Cagan headed the IDFAF procurement delegation in Europe, and he succeeded Aluf Mishne Heyman Shamir in March 1950. He contacted the French authorities with a request to procure these Mosquitos. The following months saw a prolonged negotiation in which it was revealed that while the French Ministry of Defense sympathized with Israel and was eager to help, the Ministry of Foreign Affairs preferred the Arab world, mainly for political considerations, and rejected any help to Israel. Finally, on 17 February 1951, the contract was signed and included 40 Mosquito FB.6s, 19 NF.30s, four PR.16s and three T.3s. The agreed prices were U.S.$13,000 for a serviceable trainer, U.S.$5,500 for every fighter bomber stored in a hangar, and U.S.$200 each for the rest that were parked in the open and were considered scrap. The deal included all available spare parts and the 20mm guns, but no radio sets for the whole fleet and no radars for the night fighters. Numerically, the deal implied a reinforcement of two fighter bomber squadrons, a single night fighter squadron and a reconnaissance flight. The last two were especially welcomed, and the former was to counter the REAF's bombing force, while the latter was for adding a new dimension to IDFAF capabilities in long range intelligence.

However, the Mosquitos required an overhaul before being ferried to Israel, pilots had to be trained in the new type, and accessory equipment procured elsewhere to secure the operational value. To overcome the first obstacle an overhaul depot was created at Chateaudun with a small number of Israeli engineers and technicians that supervised the work of Frenchmen, which were in fact employees of Nord who had worked on the structure, and Hispano-Suiza for the engines. The French workers were provided by both companies under a commercial contract, while the responsibility was wholly on IDFAF, headed by Freddy Ish-Shalon. While negotiating with Nord it was discovered that two overhauled Mosquito FB.6s were in that firm and both were acquired, rising the total of this mark purchased in France to 42. In March 1951, Seren John Harvey, the pilot who ferried the Mosquito PR.16 to Israel in July 1948, was sent to Chateaudun to test fly the overhauled Mosquitos, bust as no cadre of experienced Mosquito pilots existed in the

Two Consolidated PBY-5A Catalinas were acquired from the USA in 1952. They performed maritime patrol duties until phased out of service in the late fifties.(IAF)

IDFAF, a contract was signed with the British company Britavia to cover the ferrying of the Mosquitos to Israel. By the following month, the French Foreign Ministry did not sanction the export license. After lengthy negotiations it was granted, and on 11 June, the first three refurbished Mosquitos left France for Israel via Nice, Brindisi and Athens. By 7 August, twelve had been ferried and the operation continued at a slow pace; 36 were ferried by year's end, and a total of 62 ex-Armée de l'Air Mosquitos arrived in Israel without incident, except for one funny story. In accordance to IDFAF numbering sequence of the period the Mosquitos were numbered from 2101 onwards (the first, 2101, was the PR.16 that Harvey ferried from the UK). However, for the ferrying the letters IAF, for Israeli Air Force, were added before the number. During one of the ferry flights a British controller at Cyprus mistook the letters for Iraqi Air Force and thus, after the two Mosquitos left Cyprus bound for Israel, he transmitted to Iraq to expect a pair of Mosquitos. The transmission was also received in Israel, and it was feared that the Mosquitos were hijacked!. Once the misunderstanding was cleared the IAF acronym was replaced by IDFAF for Israel Force/Air Force, which now has been used up to the present.

But the Mosquito procurement did demand its victims, and three aircrews, all ex-foreign volunteers, were killed. While the Mosquitos were overhauled in France, the Canadian government approved the export of six Mosquitos to Israel, and Al Schwimmer was responsible for that project. At first a single machine was acquired and overhauled by de Havilland of Canada. Ray Kurtz, the first CO of 69 Sqn., was asked to ferry it to Israel. Kurtz, who had returned to his routine life in the U.S.,volunteered once more and flew the Mosquito to Goose Bay where he spent three weeks, due to bad weather, before his navigator had to quit so he could make it in time for his wedding. Another veteran volunteer was then also asked for one more favor, and Cy Lerner, the original navigator of the first Ballak flight and later a B-17 navigator, agreed. On 17 May 1951, the lonely Mosquito left Goose Bay and was never seen again. On the other side of the Atlantic Ocean, John Harvey, an Englishman that still served with the IDFAF in a period when most of the original foreign volunteers returned for their homes and new careers, also paid the ultimate price. During one of the test flights, Harvey's Mosquito crashed, and he was killed. The pilot that succeeded Harvey as test pilot was Hugo Marom, who had graduated from the Flying School two years earlier. In spite of the fact that the Mosquitos were in Israel, there were no pilots to operate them and they were parked at Tel-Nof and flown occasionally, to keep the systems running by Cy Feldman, IDFAF's chief test pilot at the time, Ben-Simon, Abe Nathan and Danni Shapira. Finally, the RAF agreed to accept several pilots for a conversion course and the first to be sent was Seren Abe Nathan, joined in December by six others, led by future Mosquito squadron C.O. Dani Shapira, and were Dave Orli, Moshe Ashel. Dov Erlich, Israel Lahav and Abraham Portugali. The course was at RAF Swinderley. There were some

Line up of Spitfires from 101 Squadron. After the purchase of 30 Mustangs in the USA and 25 in Sweden, the unit completly re-equipped with the type, while the Spitfires were passed to the newly activated 105 Squadron which acted as a sort of Operational Training Unit for fighter pilots. Also a short lived 107 Squadron was formed to operate these fighters, of which the majority were later sold to Burma.(IAF)

Since the end of the War of Independence the Harvards reverted to their advanced training role at the Flying School, although they continued to have a secondary role as close air support platforms.(IAF)

incidents during that course, including a British instructor that refused to fly with the Israelis, since the pilot killed in the RAF Mosquito that was downed by Wayne Peake in November 1948, was his close friend, but the group graduated and returned to Israel by March 1952 to organize the first Mosquito squadron, 110, in the reopened Hatzor air base. Wing 4, as Hatzor's parent unit was numbered, had been activated previously at Tel-Nof before moving to Hatzor. Another squadron was also activated, but the serviceability rates were low and many accidents took place. No special night fighter squadron, 109, was formed while a reconnaissance flight was activated a bit later. All types were operated by the two squadrons, and the Mosquito NF.30s, painted black and equipped with American AN/APS-4 radars, were found particularly successful in detecting ships.

The low serviceability of the period and the relative high attrition of the old fighters—Mustangs and Spitfires—resulted in the procurement of additional quantities of both types. The Mustang was generally preferred due to its longer range and the heavier loads that it could carry, but the Spitfire was also considered valuable, especially since it introduced the graduates of the Flying School to the art of fighting through the OTU. Following the acquisition of 30 Mustangs in the US, 101 Sqn. was fully converted to the type, and it was Laskov's ambition to form a second Mustang unit that led to the procurement of 25 P-51Ds from Sweden, a contract that was signed in late 1952, being ferried to Israel between December 1952 and July 1953 by RSAF pilots that were granted special permission to make the flights in their spare time and on their own responsibility. The ferrying of all the Mustangs was uneventful, but this time fictitious numbers were painted on them—although the acronym IDFAF was retained—instead of the real numbers which for the Mustangs started with 2301. The new unit was not activated following the arrival of the Swedish Mustangs, but it was a welcomed reinforcement. Concerning the Spitfires, Italy was the source for another thirty. This time the negotiations dragged on for almost 18 months, from September 1951 to early 1953, and a refurbishment center was created at Foggia, while a commercial contract was signed with Ambrosini which supplied the workers, and a IDFAF technical delegation was responsible for the work. The Spitfires were ferried to Israel without an incident and enabled the commissioning of a third fighter squadron, 107 Squadron at Ramat David, although it was short-lived, lasting for less than a year as the jet age finally arrived for the IDFAF. But even after the introduction of the first jet fighters into IDFAF inventory, the older types continued into service, and except for the Spitfire, which was placed out of service in early 1956, even participated in the Sinai Campaign of late 1956. It was during the commanding periods of Shamir and Laskov, which can be regarded in spite of diminishing budgets and controversy with the ground forces, that the air force expanded and a new base was opened. Against the two fighter squadrons and the single bombing squadron that Shamir inherited from Remez, Laskov left an IDFAF which consisted of three fighter squadrons (101, 105 and 107), two of fighter bombers, recce and night fighters (109, 110) and a single of bombers (69). It was an expansion of one hundred percent in less than three years, and this thanks to commanders who were ground officers! Moreover, during that period two other types were also procured and added to the inventory. Three consolidated PBY-5A Catalina flying boats joined the Dakotas of 103 Sqn. at Tel-Nof to support the B-17s of the neighboring 69 Sqn. (two of the fortresses had received a search radar in the nose where the gun turret used to be). The second acquisition was two Hiller 360 helicopters, in fact the first rotorcraft in IDFAF service, serving with 100 Sqn.. Obsolete and one-off types were withdrawn, the C-46s were disposed of to El Al, and the Dragon Rapides to Arkia, Israel's regional airline. Thus, by mid-1953, before Laskov left the IDFAF, the force had rationalized its force, having five active air bases and eight operational squadrons. Laskov's period as the air arm commander saw no operational activity except for the commencement of a series of long range reconnaissance flights with the Mosquito PR.16s that received a pair of underwing tanks. These reconnaissance missions were later to bring the first indications of the "Czech Deal" in late 1955, and in the meantime they supplied valuable intelligence information regarding Israel's neighboring armed forces. Laskov's follow-on was Aluf Mishne Dan Tolkovski. Before leaving Israel for his studies at Oxford he convinced Ben-Gurion that the IDFAF should be commanded by a pilot and recommended his Chief of Staff, Tolkovski, for the post. On the other hand, in his second year as IDFAF commander, fifteen Gloster Meteors were ordered. Finally, the IDFAF had entered the jet age.

Meanwhile, Egypt was suffering from internal unrest which culminated in a coup d'etat in the summer of 1952 after which King Farouk was forced to leave Egypt and abdicate in favor of his infant son, but one year later in June 1953, the military ousted him and took over the full control of the country, which became a republic. Thus, the Royal Egyptian Air Force became the Egyptian Air Force. Rapidly Egypt commenced a nationalist/anti-colonialist policy that resulted in the Suez Canal nationalization in 1954, closing the British bases in the area, a violent anti-Israeli campaign, and an approach to the eastern-bloc world exemplified through the up to then largest arms contract ever signed by the Egyptians, which became known as the "Czech Deal." Meanwhile, in Israel the IDFAF was getting nearer to touching its first jet fighters. In 1952 negotiations with Gloster commenced; two pilots, Menachem Bar and Benjamin Peled, and a technical delegation were sent to the UK to evaluate the Meteor and its Rolls Royce Derwent engines, and they stayed there for four weeks. Although the Meteor was already obsolete in the days when the F-86 Sabres and MiG-15s were dueling over the Korean skies, it was an excellent introductory jet fighter, not just because of its stable flight characteristics, but also because it was the only one available to Israel. A contract was signed for 11 Meteor F.8s at a unit cost of 36,250 Sterling Pounds, and four Meteor T.7s, each costing 34,250 Pounds. The two-seat Meteors were purely trainers; they carried no weapons at all, and thus had no operational value, while the single seat fighters were equipped with four Hispano 20mm guns with 250 rpg. In addition, they were modified to carry eight 5-in HVAR air-to-ground rockets under its wings, while it had the capability to carry a belly auxiliary tank as well. The first two Meteors arrived at Ramat David air base on 17 June 1953, flown by British ferry pilots, but were accepted in a formal ceremony which was attended by the Prime Minister and Minister of, David Ben-Gurion, who named them (both were T.7s) as Sofa (Storm) and Sa'ar (Assault). Two months later the gleaming silver F.8s arrived, and 117 Sqn., also known as the "First Jet Squadron", commenced its working-up period. First commanding officer was Rav Seren Menachem Bar, and the unit initially concentrated on two issues: become operational and convert as many pilots as possible to jets. The first five to attend a conversion course which was given by Bar and Peled were Moti Hod, Aharon Yoeli, Joe Allon, Jacob "Yak" Nevo and Danni Shapira. The last Meteors of that initial order arrived on 21 January 1954, by which time 117 Sqn. was operational.

Meanwhile, due to the budget restrictions of the period the first export of refurbished military aircraft by Israel took place. In May 1951 during his visit to the U.S., David Ben-Gurion invited Al Schwimmer, who had returned to the U.S. and formed a small aircraft maintenance and refurbishment company in California, to transfer his activities to Israel. The outcome of that invitation was Bedek Aviation (Bedek in Hebrew means Maintenance), which commenced operations during 1953 at Lod International Airport east of Tel-Aviv. Both installations changed titles in the coming years: Bedek Aviation became Israel Aircraft Industries (IAI), although its division which provides refurbishment and maintenance services is still called Bedek Division, while Lod became Ben Gurion International Airport after the name of the Israeli leader who died in late 1973 at the age of 89. During its first period of existence Bedek Aviation mainly supplied maintenance services to the IDFAF, inaugurating the close ties that continue up the present, but also sought export contracts, and one of these was the sale of 30 refurbished Spitfires to Burma. The Spitfires were those withdrawn from use when 107

For a short time the B-17Gs of 69 Squadron received some indigenous chin turrets, as seen in this rare picture of an uncamouflaged machine, the threesome acquired the serials 1601, 1602 and 1603. (IAF)

Squadron at Ramat-David disbanded in mid-1954, although 105 Sqn. continued to operate the beloved fighter as an OTU until replaced by Mustangs early in 1956. The Burmese government needed the Spitfires as fast as possible because of the internal situation of the country and thus demanded that they would be flown to Burma instead of delivered in crates by sea. The agreement was that Bedek Aviation would fly the refurbished Spitfires to Cyprus and hence a private company, Air Services of Croydon, under contract to the Burmese government, would take control of the operation and ferry the fighters all over to Burma. However, during one of the flights three Spitfires had to land at Beirut, Lebanon, where the pilots were arrested and the aircraft interned. As a result, Air Services abandoned the ferrying operation and Burma threatened to cancel the order for the remaining 16 Spitfires. Pressed to the wall, Bedek Aviation took under its responsibility the ferrying operation and called for the help of three IDFAF foreign volunteers, the Americans Leo Garner and Sony Benting and the British Gordon Levett, plus a woman pilot, Jackie Mogridge, who previously ferried the Spitfires under a personal contract with Air Services. The remaining 16 Spitfires were delivered in four flights that could hardly be described as uneventful, but which escapes the scope of this narrative.

During that time period there was a heavy attrition of aircraft in training and in a smaller part to flying discipline, both being slowly corrected, and thus in spite of the availability of the Meteor-equipped 117 Sqn., the IDFAF still relied heavily on piston-engined fighters, and that meant that to cover losses and to create a reasonable reserve in case of emergency more had to be acquired. In November 1953, the procurement delegation in Europe commenced negotiations with Mr R.A. Short, an Englishman who bought 73 almost new Mosquito TR.33s from the Royal Navy, and a contract for 20 was signed, but as the British government only sanctioned the deal in June 1954, only 14 were eventually bought, and after refurbishment by Eagle Aviation were ferried to Israel between October 1954 and June 1955. Finally, another 30 ex-Italian Air Force Mustangs were procured during 1955, but only a fraction, about two thirds, entered IDFAF service, while the remaining became a source of spare parts. All these aircraft were destined to prop up the strength of established units and to create an adequate attrition reserve. Yet another surprise was the approval by the British government of the sale of seven Meteor FR.9s in March 1954. These aircraft had been used by Flight Refueling Ltd for air-to-air refueling tests and had to be refurbished by the manufacturer before ferrying them to Israel between January and May 1955. Just as with the Mosquitos and Mustangs, the additional Meteors augmented 117 Sqn. at Ramat David, and no new units were formed during that period except for a specialized reconnaissance flight with Mosquito PR.16s that years later would become 115 Sqn..

The growing threat of a renewal of hostilities, coupled with the fact that the IDFAF finally had a long range high altitude reconnaissance platform, led to an intensive campaign of operational sorties to obtain first hand intelligence about Israel's surrounding Arab states. In June 1954 a recce flight was commissioned at Hatzor under the command of Azriel Spivak, a graduate of the Palmach's Flying Platoon. The silver unarmed Mosquito PR.16s were lightened by deleting all unnecessary equipment and usually flew at 30/35000 ft. Two external underwing auxiliary fuel tanks were added for long range missions and gave the recce Mosquitos an endurance of up to five hours. The usual targets were Yehuda and Shomrom (in the West Bank under Jordanian control) and Sinai Peninsula, while among the more risky long range operations were flights to Siwa in the Western Desert in Egypt where Israeli agents were held by the Egyptians, Mersa Matruch naval base in Egypt, H-3 air base in Iraq, El-Maza air base in Syria, and Luxor air base in the Nile Valley in Egypt. Throughout the period in which the reconnaissance Mosquitos wandered around the hostile skies of the Middle East, not a single aircraft was lost, although many mechanical failures were encountered and as a result many missions were aborted. Nevertheless, the effort of its crews paid handsomely and the intelligence gathered during these flights was of great value.

Close-up of a 69 Squadron crew shortly before a mission, just after the end of the War of Independence, still with the old serial (21, 22 and 23), and carrying bombs in underwing racks.(IAF)

Another aircraft was also involved these days in long range surveillance missions. The B-17 Flying Fortress was not perhaps the ideal machine for the task, but its endurance and the fact that two were equipped with surface search radars made them the perfect choice for maritime patrol. On the other hand, the third B-17 retained the classic configuration of three gun turrets, while each of the radar-equipped B-17s only had two turrets. By late 1954 the three Flying Fortresses were part of 103 Sqn., as 69 Sqn. had deactivated as an economy measure, but were very little used; but then, in September 1954 the long range and endurance of the B-17 was brought to good use when a merchant ship went missing. As part of the armistice agreements of 1949 Egypt had to allow the free passage of Israeli or Third Country merchant shipping, which carried goods to or from Israel, through the Suez Canal. Israel put the Egyptians to the test several times, but to no avail, and in spite of the U.N. intervention the Egyptians stuck to their particular views. In fact, part of the general Arab view was that Israel's economical troubles would weaken it to such an extent that a second war might not be eventually needed, so it was decided to test the Egyptians with an Israeli ship. Bat Galim (Daughter of the Waves) was a small merchant vessel that left Massawa in Ethiopia with a load of wood, leather and meat to Haifa. Radio contact with the ship was lost on 28 September 1954 and a B-17 was sent to search for the missing ship. The captain of the Flying Fortress was Rav Seren MIsha Kenner, who was the last commander of 69 Sqn. before its deactivation and its four-engined bombers transferred to 103 Sqn.. The last known position of the ship was some 250 km south of the strait of Tiran, and the B-17 was equipped with Tokyo tanks that added another four hours of endurance, and thus the search track was made so that it would fly south of that point and search to the north. The crew was accompanied by an Israeli Navy officer who was supposed to identify the missing ship and for that purpose the large aircraft flew very low with the crew virtually reading each merchant ship's name. Despite the effort Bat Galim was not found until the Egyptian port of Sharm el Sheikh in Sinai, just on the entrance to the Gulf of Suez, was reached, and the ship was positively identified beside a dock. The Egyptians arrested the ten Israeli seamen and released the ship and its crew only four months later. The total flying time of the B-17 on that mission was eleven hours, a remarkable achievement, although it was not the longest B-17 flight recorded in IDFAF service as in another operational sortie a total of 13 hours was logged. However, the search for Bat Galim was the temporary swan song of the Flying Fortress in Israeli service. The growing maintenance costs to keep them in the air, and the appearance of jet fighters in the area caused the temporary retirement of the four-engine bombers. The three B-17s were flown to Lod International Airport where Bedek Aviation maintained them as a reserve in the event of an emergency, and indeed in October 1956, 69 Squadron was reactivated and the old B-17s went to war again, probably as the last Flying Fortresses tasked with a first line operational role on earth.

Among the economical blockades to Israel by Egypt at the time was not only the closing of the Suez Canal to Israeli or Third Country ships trading with Israel, but also the closing of the Strait of Tiran, which connects the Gulf of Aqaba to the Red Sea. This measure rendered the port of Eilat useless, meaning that Israeli ships had to sail around the continent to reach East Africa, Asia and the Far East. The fact that Israel was backed by international law was to no avail, and the Egyptians positioned a battery of shore guns at Ras Nazrani, opposite to the island of Tiran, which was officially Saudi Arabian territory but was in fact held by the Egyptians, and searched ships which sailed through the straits—an international route—for Israeli goods or goods that were destined to Israel. It was an unbearable situation, and Israel soon realized that it could not rely on the international community, and if it desired to open the Strait of Tiran to Israeli shipping then a military move was the only possible solution.

The IDF's General Staff commenced planning an operation to capture the eastern coast of the Sinai Peninsula, from Eilat down to Sharm el Sheikh, in order to open the Strait of Tiran to Israeli shipping. No road of any kind existed in those days between Eilat and Sharm el Sheikh, while the present day road was constructed by Israel only after the Six Day War of 1967. The Egyptians supplied their garrisons at Sharm El Sheikh, Ras Nazrani, and the Island of Tiran by ships from the western coast of the Gulf of Suez. In spite of the availability of aerial photography provided by the Mosquito PR.16s, the real character of the local terrain was unknown, and it was clear that without a firm and stable connection—which means a land route—between Eilat and Sharm el Sheikh, any such operation was doomed as the Israeli forces that might capture the area would be isolated some 200 km away from the nearest Israeli point, and only 70 km away from the coast of mainland Egypt. Thus, it was well understood that a small unit of surveyors had to penetrate deep into Egyptian territory and survey the more questionable part of the planned land route. Six men, including five officers, volunteered for the mission, which was named Operation Yarcon after the small river just north of Tel-Aviv and south of Sde-Dov. The unit was to receive the full cooperation of the IDFAF, with the 100 and 103 Sqns. bearing the main burden. The Dakotas of the latter were planned to drop water cans to the unit at predetermined locations and also to serve as communications relays under a predetermined schedule. The Piper Cubs of 100 Sqn. were to stand at readiness for any possible evacuation of the unit. Also, the Dakotas were seen as a possible evacuation vehicle, and on 28 May 1955, 103 Sqn. received an order to commence immediate training on takeoffs and landings from short, unprepared strips of no more than 250 meters long. To locate a suitable place in the area of operations a single Piper Cub flown by Seren Hanuch "Herbert" Keret and accompanied by the commander of the surveyors, Sgan Aluf Asher Levy (CO of the 51st Regiment of the Givatti Brigade), was sent to the area. Although the possible landing strips were located, the Piper Cub's Lycoming engine suffered from overheating, not unexpected in the hot summer days of the Sinai Desert, but it made it back to Eilat.

On 9 June 1955, a small fishing boat named "Eilat" and operated by the Israeli Navy landed the small unit of surveyors on the shore of Dahav on the western coast of the Gulf of Aqaba, just opposite to the area of Al Maqnah on the western coast of the same gulf where the Bar Giora incident took place the previous year. The surveyors commenced immediately to explore the route to the north, towards Eilat, and a Dakota flown by Yoseph Offer and Ori Yaffe served as a communications relay that day while the next day, 10 June, four water cans were dropped from a low level by a Dakota, although only one was recovered intact. But this time it became clear that an Egyptian unit was following the tracks of the surveyors, and it was decided to evacuate them the next day. As early as 9 June six Piper Cubs were flown to Eilat and waited there for a possible evacuation. However, to remove any suspicion that forces of any kind were concentrated at Eilat, five of the light-planes were dispersed in the area, while the sixth, equipped with a stretcher for medical evacuation, was at immediate readiness at Eilat. In the very early morning hours of 11 June, possibly to avoid overheating their Lycoming engines, the six Piper Cubs left Eilat in two formations of three each. The leader of the first formation was Hanuch Keret, and the other pilots were Moshe Aran, Shlomo Brosh, Itschak Zamir, Eliezer Levinson and Eli Atar. Once more the liaison pilots of 100 Sqn. demonstrated their skills, and the six surveyors were evacuated to Eilat. For various political reasons the planned operation did not materialize, but the information gained during Operation Yarcom served the 9th Brigade, which followed the same route in its thrust towards Sharm el Sheikh during the Sinai Campaign.

Despite the fact that Israel did not open by force the straits of Tiran blocked by Egypt, tension in the area continued to climb and the intensity and scale of the local skirmishes grew steadily to such an extent that air power had to be used, although in a very limited way, culminating in the first all-jet air combat between Israel Meteors and Egyptian Vampires. As the tension on the Israeli-Egyptian border grew and the EAF intruded into Israeli air space from time to time, a detachment of Meteors was established at Hatzor, some 30 km from the Egyptian border, as Ramat-David in northern Israel was 140 km away from that same border. The Meteors' deployment at Hatzor, where only the Mosquitos of 109 and 110 Sqns. were resident at the time, reduced the reaction time to several minutes, and thus El-Arish based EAF fighters would have much less leisure time intruding into Israel. The first opportunity to achieve the first victory for the Meteor in IDFAF service fell to Seren Aharon Yoelli, one of the two flight commanders of 117 Sqn., in July 1954. A pair of Meteors were scrambled to intercept an Egyptian incursion, but only the leader, Seren Yoelli, took off, as his wingman had a mechanical failure. Nevertheless, Yoelli pressed on, but the Egyptian Hawker Fury, probably warned by Egyptian ground control, turned back to the safety of Sinai. The frustrated Yoelli ignored all orders and crossed the border to buzz El Arish air base at a very low level as the EAF Fury came in to land. Not only did he miss his prey, but he was also formally reprimanded by his base commander, Ezer Weizman, for crossing the border against all orders and was fined 25 Israeli Pounds, a respectable quantity in those days.

Line-up of Mosquitos from 110 Squadron. These machines would be replaced by an advanced French design, the Sud Vatour.

Chapter 8: The First Jet Kills

METEOR VERSUS VAMPIRE

During the winter of 1954/1955, the Egyptian Air Force (EAF) piston-engined Hawker Furies based at El Arish were replaced by jet powered Vampires, and although its overall performances were still inferior to the Meteor F.8s and FR.9s which then were the IDFAF top fighters, they were more agile, and its fastest top speed compared to the Fury meant that it stayed less time during the incursions into Israeli air space, which were usually less than 30 km deep. By the coming summer the tension on the Egyptian border had reached new peaks, and the Meteors at Hatzor air base waited for the inevitable clash. On August 22 a local skirmish developed when an Egyptian strongpoint opened fire on a routine IDF patrol. The two-phase retaliation commenced with an immediate attack on that specific strongpoint, which was duly captured, and some 25 Egyptian soldiers were taken POW.

It was Monday, 29 August, when at last the Meteors managed to jump the Egyptian Vampires before they cleared the Israeli territory. It was the first jet combat ever recorded in the Middle East, and its results were not conclusive. The pair of IDFAF Meteors intercepted four EAF Vampires over the Israeli settlement of En Ha'shlosha and Kisofin, just east of the Gaza strip, and a short combat, which lasted for about a minute, developed. The Israeli leader managed to acquire one of the EAF Vampires and fired a long burst from his four 20mm guns. He noticed some hits on the Egyptian fighter, and then some pieces flew-off and the Vampire disappeared from sight to the west, towards the Gaza strip, between the clouds while diving. The other three Vampires abandoned the fight for the safety of Sinai, and the Meteors returned to Hatzor with mixed feelings: the Vampires had been expelled out of Israel's air space for the first time, but the battle was not conclusive and no victories were officially achieved. The next day Egypt admitted the loss of an EAF aircraft during a "training mission" and it was alleged that the lost aircraft was in fact the Vampire that was at least damaged by the leader of the Meteor section, but the claim was never proved and 117 Squadron had to be patient for three more days to achieve its maiden kills.

On Wednesday, 31 August, the pair of Meteors which scrambled from Hatzor to intercept an intruding formation could not reach the area in time and the four Vampires managed to clear Israeli air space. During the following night, 31 August/1 September, the long awaited second phase retaliation operation against the Khan Yunis police station was launched, and it was a very successful combat for the IDF; the building was blown up as planned, and the Egyptians suffered a large number of casualties, 72 killed and 58 injured. The IDF forces lost one soldier killed and eleven injured. The next morning, Thursday, the EAF Vampire incursion was deeper than ever, and it was also reported from kibbutz Carmin, just north of Gaza Strip, that one of the Vampires fired a short burst aimed at the settlement. It was probably a show of force by the EAF, a strange retaliation to the previous night battle, but probably aimed to inform the Israelis that no matter how successful their retaliation operations might be, the Egyptian forces were still strong and capable. But this was to prove the opposite, and the demonstration of force was exactly what the 117 Sqn. pilots at Hatzor were waiting for.

The scramble horn was heard just before 0700 hrs, and the first to reach the waiting Meteors were not the pair of pilots on readiness at the time but another pair: Seren Aharon Yoelli, who had celebrated his 26th birthday the previous day, and his wingman, Seren Yehuash "Chato" Tsidon. It was true that they were not the pair of pilots at readiness, but they were the pair who ran faster! Those were still the early, adventurous days within the IDFAF, without too many procedures applied, or, as Ran Peker once saide, a seven-kill ace and one of the greatest fighter leaders in the history of the Israeli Air Force, "there are no sentiments in a war, nor in a state of war in which we lived for so many years." Thus, the fastest ones to sprint to the aircraft were those to take-off and engage the EAF Vampires, with no sentiments at all. The surprise that awaited

A 69 Squadron B-17G Flying Fortress equipped with its original Bendix chin turret, displaying some of the armament it could carry.(IAF)

First jets for the Israeli Air Force, the ceremony after the arrival of the two Meteor T.7 two seat trainers which marked the commissioning of number 117 Squadron in 1953, nicknamed the "First Jet Fighter Squadron"*. This unit has lived since then a distinguished career, also flying Mirages, Neshers, F-16A/Bs and actually F-16C/Ds, scoring dozens of air-to-air victories. (IAF Magazine)*

Yoelli and Tsidon in their respective cockpits was that the gunsights had been set for night flying, and illumination of the diamonds which formed the pipper was the least possible and the range pegged at 100 meters; but they had no time to change it to daytime status and the two fighters roared off immediately. The four Vampires that had intruded Israeli air space as far north as Ashkelon, some 20 km from the Egyptian border and only about ten km away from Hatzor, were already on their way back home, benefiting from very cloudy skies. Yoelli and Tsidon were at an altitude of some 5,000 ft when "Chato" called "two Vampires, three o'clock low!" Yoelli, flying Meteor FR.9 number 36, dived at the Egyptian wingman, with Tsidon covering his tail, and opened fire at a range of about 150 meters. Yoelli did not trust on his gunsight, fixed for night operations, and used the fact that every fifth round was a tracer, again for night activity. He saw that his rounds were finding their mark, and then the Vampire, number 1569, dived into the ground trailing black smoke and crashed in a corn field near kibbutz Carmis some five kilometers north of the Egyptian border. Its pilot made no attempt to abandon his aircraft and was killed. Then Yoelli turned his attention to the Egyptian leader. While the Egyptian wingman probably never saw his attacker before the bullets struck him, the leader was already well aware of the presence of the Israeli fighters and commenced several rolls at an altitude of some 1500 ft. Yoelli just waited for the EAF pilot to finish his aerobatics display and then closed the range to 200 meters and opened fire. He saw how his rounds hit the Vampire's cockpit, and the aircraft, serial number 1567, crashed some two kilometers away from the place where the first Vampire crashed, its pilot being killed also. The Egyptians admitted the loss of two EAF aircraft in a "mid-air collision."

At 0720 the pair landed back at Hatzor. It was a great birthday present for Seren Aharon Yoelli. Years later and after a brilliant career, he retired from the Chel Ha'avir with the rank of Aluf Mishne, the same rank that his wingman for that mission, Yehuash "Chato" Tsidon, had. For 117 Squadron these were the first victories, and the First Jet Fighter Squadron had to wait for almost eleven years before its third official victory was recorded, when a Syrian MiG-21 was downed by Seren Yehuda Coren in a Mirage IIICJ, and from then on it became the top-scoring IDFAF fighter unit, getting

Gloster Meteor T.7 dual trainer number 17. (IAF Magazine)

A pair of 117 Squadron Meteor T.7s flying over Izrael Valley. The hebrew letter before the idividual serial, Eemoon, *means Training. (IAF Magazine).*

The Dassault Ouragan was the first French fighter acquired by Chel Ha'avir and although less performing than the MiG-15, it proved a sturdy and stable weapons platform. Illustrated is number 59 of number 113 Squadron, a few of which sported a large sharkmouth on the nose.(IAF Magazine)

kills in Mirages, Neshers and F-16 Fighting Falcons. For the Israeli Air Force it was also the first jet kills, and the incident was a major boost to its morale, but it was exactly the same month when an event which was to change the Middle East military balance took place. The "Czech Deal", announced by Egypt's President Gammal Abed al Nasser in September 1955, marked the rise of the USSR to the status of the main arms supplier to most Arab states.

PREPARING FOR WAR

On 27 September 1955, Egypt's President announced that he had taken the first steps in his wish to achieve a numerical and technological superiority over his hated enemy, Israel. The signing of the "Czech Deal" meant the supply of all kinds of advanced weaponry never seen before in the Middle East countries, traditionally supplied with hand-me-downs by the Western powers. The deal was in fact signed with the USSR, but Czechoslovakia figured in only for political and commercial reasons (also some equipment of Czech origin was delivered). Nasser was happy with the deal, since no political demands were placed by the Soviets. In fact, it had been almost a barter, Soviet weapons for Egyptian cotton and rice. Egyptian military went to the USSR to learn to operate the new equipment, while Soviet and Eastern Bloc advisors traveled to Egypt to supervise the assembly and training. Other Arab countries would soon follow the Egyptian example.

The "Czech Deal" totally changed history almost immediately. The U.S. and Great Britain, facing the ultimate loss of influence with the leading Arab country and fearing that the new arms would give Nasser the opportunity to attack Israel, pressured the Egyptian government to cancel the deal, but to no avail. Nasser was determined to achieve his goals, and it became a matter of prestige since the aim of the "Czech Deal" was to end the Westerner's control over Egyptian policy, so how could Nasser surrender to the U.S. and Great Britain?. Moreover, it was a critical period of the Cold War and the Soviet achievement caused real concern in Washington, and in spite of the pressure applied the deal was concluded. The feeling in Israel and the Western countries was that a war was inevitable.

But what was the Czech Deal all about and why was it so dangerous to the existence of Israel? Both parts of this question can not be answered without speculation. Even at the present time the specific quantities of weapons supplied to Egypt have never formally been released, and thus all data available is either based on intelligence reports, which tend to exaggerate, or press reports, which tend to be inaccurate. The second part of the question can only be speculated upon since before the armed forces of Egypt absorbed the influx of Soviet weaponry, Nasser's wishes to attack Israel were frustrated when in late 1956 British and French aircraft wrecked most of Egypt's shiny new air force while on the ground. The deal not only included aircraft, but every type of equipment for both the army and navy up to the level of personal weapons. However, all these important items are beyond the scope of this feature. The aeronautical part of the deal was quite impressive, and the aircraft involved were MiG-15 jet fighters, Il-28 jet bombers, Il-14 piston-engined transports and Yak-11 basic trainers, all replacing similar types of British and American design. Thus, these supplies not only provided Nasser with a qualitative edge—the Il-28s were considered a threat to Israeli cities—it also gave him the much desired numerical superiority. Again, it should be stressed that no reliable figures exist, only estimations. The guessing about the number of MiG-15s and MiG-15UTI two-seat conversion trainers has varied between 86 and 200. Evidently, it was somewhere between these two figures and probably close to the lower estimation. Concerning the number of Il-28s, it varied between 39 and 70, and once more the truth probably lies somewhere in between these figures. A quite realistic Israeli estimation of the period claimed that the total number was 100 MiG-15s and 48 Il-28s, although David Ben-Gurion, who returned from his political retirement at kibbutz Sde Boker to the post of Minister of Minister of, replacing Pinchas Lavon in February 1955, and Prime Minister, replacing Moshe Sharet nine months later, stated that Egypt was to receive 200 MiG-15s and 60 Il-28s. This statement was surely exaggerated on political grounds: to turn the attention of the free world to the desperate situation of Israel facing its main enemy re-equipping with numerous modern weapons.

By the end of 1955 Israel had three options, or a mix of the three: to procure equivalent weapons, to launch a pre-emptive attack, or to wait for an Egyptian attack which was not expected before the new Soviet weapons would be fully absorbed into operational service. Although it takes years to fulfill a military contract and turn the status of the new equipment to operational, the supply of the Soviet arms commenced very quickly and at an alarming rate. The first MiG-15s arrived at Alexandria's harbor in November 1955, and the first Il-28s arrived several weeks later. The rate of deliveries was so fast that by the spring of 1956 the EAF had an OCU with MiG-15s at Kabrit in the Suez Canal Zone, while two MiG-15 squadrons and a single Il-28 squadron were working up in an air base around Cairo. The numerical strength of that force was possibly some 50-60 MiG-15s and 10-20 Il-28s. Thus, although Israel's procurement efforts bore fruit, it was feared that the rapid expansion of the EAF would provide Egypt with total air superiority by the summer of 1956.

The only fighters in the market roughly equivalent to the Soviet MiG-15 were the Sabre, Mystère, Hunter and SAAB J-29. The British Hunter was probably the best, but it was not considered for the IDFAF, which evaluated the French Mystère while awaiting the outcome of the negotiations concerning the procurement of Sabres. However, it was the purchase of the Dassault Ouragan that marked the start of the tight and warm relations between Israel and France which lasted for no less than fifteen years until President De Gaulle embargoed the 50 Dassault Mirage 5s, already fully paid for, in June 1967.

In July 1952, following American and British pressure, France turned down an Israeli request to acquire Ouragans. By then this aircraft, the most modern fighter in the French service, was already obsolete. An unexpected surprise was the positive attitude of the Swedish company SAAB, which informed Israel in November 1953 that it was willing to sell its excellent J-29 jet fighter if the Swedish Government would approve such a deal. It was quite clear that the government sympathized with Israel as already exemplified by its willingness to sell 25 ex-Svenska Flygvapen Mustangs, and gave a positive answer to the Israeli request for 18-24 fighters, depending on the financial credit provided. Almost two years before Nasser announced the procurement of MiG-15s, Israel could acquire an equivalent fighter!

But the deal fell through as the French, at last, offered Israel jet fighters. The aircraft in question was not the Ouragan, but a straightforward derivative, the Mystère II, in which the Ouragan's fuselage was mated to a new swept wing, a new tail section and a new engine, the SNECMA Atar 101 with 2800 kg of thrust. For the first time in its history the IDFAF was able to evaluate two types of fighters for possible acquisition and a technician was sent to Europe. Since no fast jet test pilots existed in Israel, the only pilot in that delegation was Seren Benjamin Peled. He flew the two types, and with the Mystère II became the first Israeli pilot to break the sound barrier in a shallow dive. After returning to Israel the decision fell in favor of the French fighter; the Swedish J-29 was an excellent design, but it was a dedicated short range interceptor (in its initial variant), while the Mystère II had longer range and was under development, which implied that it incorporated newer technologies and could be tailored to specific Israeli specifications. Such a decision implied that the IDFAF would have to wait longer for its new swept wing jet fighter. Still, it was early 1954, almost two years before Nasser announced the Czech Deal, and there was not any urgency.

Dassault Ouragan just delivered from France with the civilian registration 4X-FRB still applied. (IAF Magazine)

The first Chel Ha'avir jet fighter pilots: second from right is Menachem Bar, fourth is Danni Shappira, fifth Joe Allon, sixth Binyamin Peled and seventh Moti Hod. Most arrived to high posts within the Israeli Defence Forces.(IAF Magazine)

As the Israeli Air Force decided to procure the Mystère II, yet another unexpected offer came when Canada offered Israel the Canadair Sabre with Orenda 10 engines. On technical grounds the Sabre was totally superior to the Mystère II, but the wishes of the Canadian government to export Canadian products were stopped by the U.S., which banned the sale and also turned down an Israeli request for 24 F-86Fs to be procured directly from North American for $12 million. The Mystère II remained the only valid option, and on 19 August 1954, a contract was signed with Dassault, with first deliveries targeted for late 1955. In October 1954, Rav Seren Benjamin Peled, the intended Mystère II squadron commander, was sent to France in company with a small number of pilots for conversion to the new fighter. However, the Mystère II's development was prolonged and problematic to such an extent that the Armée de l'Air decided to opt for a totally different version with a new fuselage, a modified wing, a new tail section and the proven Rolls-Royce Tay engine produced in France by Hispano-Suiza. Although named Mystère IVA, it was a completely new design which bore only superficial resemblance to its predecessor. This change left the IDFAF two options: to proceed with the planned procurement of the Mystère II or wait for the Mystère IVA. Dan Tolkowski decided to wait for the new model, as the new EAF equipment would not become operational until the summer of 1956, while as an immediate solution to acquire the Ouragan, which was inferior to the MiG-15 but still a valuable and capable fighter. The announcement of the Czech Deal pulled off some of the American and British pressure on France for not selling jet fighters to Israel, while the inability to supply the Mystère II on schedule resulted in a positive reaction to Israel's request for immediate delivery. A contract for 12 Ouragans, diverted from Armée de l'Air orders, was quickly signed, and on 6 October 1955, five Israeli pilots, led by a French pilot, ferried the first six Ouragans via Rome and Athens to Israel. A similar delivery flight with the same number of aircraft was made on 12 October, and thus, by the time that the first MiG-15s arrived in crates to Alexandria harbor in November 1955, the IDFAF had doubled its number of jet fighters per squadron. The Ouragan unit, 113 Squadron, was formed at Hatzor, where one of the two resident Mosquito squadrons was disbanded, under the command of 27-year-old Rav Seren Benjamin Peled.

Twelve Ouragans was a small number for an operational squadron, and a more acceptable level was achieved toward the end of 1955 when another 12 Ouragans were procured, also diverted from Armée de l'Air contracts, and flown to Israel on 23 November. The arrival of the Ouragans resulted in the camouflaging of the Meteors in a pattern of blue and brown upwards and light sky blue downwards, while the Ouragans retained their sliver livery—some even had a spectacular shark mouth painted on the nose—as Chel Ha'avir's new interceptors. The Ouragan was not in the same class as the MiG-15, but it was an excellent introduction to French equipment and it had some innovations, including provision for a g-suit which enabled its pilots to sustain up to seven g's without blacking out, against the maximum of four in the Meteor which lacked provision for this equipment. The Ouragan also had electrically-operated trim tabs and a SNCASO ejection seat, which, although inferior to the Martin-Baker fitted in the Meteor, gave the pilot a certain degree of confidence. As happened with the Meteors the first

After the Ouragan came the much improved Dassault Mystère IVA, here one of them with the delivery serial 6649 landing at Hatzor air base after its ferry flight from France.

goal of the new squadron was to convert as many pilots as possible. Among those converted during these few months of operations were Moti Hod, David Ivri and Amos Lapidot, all to become future IDFAF commanders; Ran Peker, a superb pilot and a seven-kill ace; Aharon "Yalo" Shavit, with three kills; Rafael Har-Lev; Sabtai Gilboa, Zoric Lev, Shaike Barreket; and David Kishon, who was credited with the first victory in the new French fighter.

The tension between Israel and Egypt grew from day to day during the winter of 1955/1956. The infusion of Soviet Equipment gave Nasser the confidence to announce, in September 1955, that any shipping sailing through the Strait of Tiran should inform the Egyptians 72 hours in advance. The same applied to air traffic, and as a consequence El-Al had to cancel its flights to South Africa and Eilat harbor was deserted. Egypt had no legitimate right to close the strait of Tiran since international law states that any strait that forms an entrance to a gulf in which the shores are of more than one country is defined as international waters. It was a casus belli by all means, but the Israeli leaders were not prepared yet for an all-out war and tried to solve the problem through the U.N. and the Western powers. Meanwhile, the scale of the retaliation operations grew to regimental levels. The Israeli Air Force continued to support these operations with the Piper Cubs of 100 Squadron. In October 1955 three Piper Cubs, operating from Be'er Sheva, supplied communications relay and casevac duties for the unit that attacked an Egyptian outpost at Kuntila, some 50 km northwest of Eilat. During the night and the following dawn the Pipers acted as communication relays, and in the morning, to help the retreating force, two of the light-planes evacuated two wounded soldiers. Another two soldiers were killed against the Egyptian losses, which included 12 killed, six wounded and 29 taken POW. In spite of the fact that the retreating Israeli force was accompanied by these POWs, the EAF tried to attack the column in the early morning hours, although the appearance of IDFAF fighters forced the Egyptian pilots to abandon their mission.

Meanwhile, the diplomatic efforts concerning the closure of the Strait of Tiran came to nothing, and by November 1955 planning had commenced for Operation Omer. Its intention was to open by sheer force the Strait of Tiran. It was to be an assault operation with the paratroopers capturing Sharm el Sheikh, while another force was to follow the land route from Eilat which had been partly surveyed during Operation Yarcom. One of the IDFAF's missions, as defined in the planning of Operation Omer, was to destroy the EAF in its bases as a preemptive move. Operation Omer was never approved by the Israeli government, and thus, later in the same month, the idea of the preemptive strike was reactivated by Dan Tolkovski. However, he was still convinced that the EAF would not attain operational status with its new Soviet equipment until the late spring of 1956, and thus he stated that the preemptive strike was dependent on the supply of F-86s or Mystères to Israel. If such fighters were to reach Israel by July or August 1956, then the strike would not be promoted, but as one shall see, such fighters were supplied to Israel even earlier.

Meteor NF.13s from 119 Squadron. Note the radar warning and homing system on the first aircraft fuselage cone.(IAF Magazine)

Photographed in July 1954 is a 117 Sqn Meteor FR.9 flying over northern Israel.(IAF Magazine)

Anyway, the success of such a preemptive strike can be highly suspected considering the mostly outdated equipment of the Chel Ha'avir at that time, and it only shows the desperate situation of Israel, which faced a real threat to its virtual existence only seven years after its creation. By the end of 1955 the sharp end of the IDFAF consisted of two wings which included an Ouragan and a Mosquito at Hatzor, and single Meteor, Mustang and Spitfire squadrons at Ramat David. The Harvards of the Flying School at Tel Nof were also considered for emergency use in spite of their obsolescence and vulnerability. The supporting elements included the transport and maritime patrol aircraft of 103 Squadron at Tel Nof, and the light-planes of 100 Squadron at Ramle, just adjacent to the IDFAF staff camp. 103 Sqn. added a single Noratlas transport to its inventory by the end of 1955. The procurement of the Noratlas was against the will of the Israeli Air Force, which preferred to spend its limited procurement budget on fighters, and emerged as a direct result of the Ouragan deal. Three of the French transports were initially procured, the first arriving in Israel before the end of 1955 and the other two during early 1956, augmenting the venerable Dakotas with 103 Sqn.. Although the twin-boomed freighter was highly appreciated and gave sterling service during many years, they never replaced the Dakota, which in 1996 still flies in IDFAF colors, while the Noratlas was phased out of service in late 1978!

The Flying School's Steramans were also considered for emergency use but only for internal liaison and CO's transportation, while the three B-17s, as well as a number of Mustangs, Mosquitos and Spitfires, were held in storage at Bedek Aviation at Lod, and could be activated in a time of crisis. It seems that this force was insufficient, both in quantity and quality, for a successful preemptive attack, although speculating the outcome of such an attack, weather successful or not, is beyond the scope of this feature.

The early months of 1956 were relatively quiet, and only towards the spring did the activity return to its "normal" level. The assumption that the Egyptians required at least several months to absorb the influx of new Soviet equipment proved correct, but the EAF continued to operate the Vampires from El Arish, only this time they were backed by MiG-15s based at Kabrit in the Suez Canal. By April the Ouragans were nearing their first six months of service as IDFAF's most advanced frontline fighter. That month was to be their last in such a position, and it was only fair that the maiden victory of a French fighter in Israeli service was achieved by an Ouragan just before the type lost its silver finish to the common blue/brown and light blue sky camouflage in favor of the newly arrived shining silver Mystères.

The EAF Vampires from El Arish continued to intrude quite often into the Negev, and 113 Squadron at Hatzor maintained a pair of Ouragans on QRA from dawn to dusk. On 12 April 1956, just as another uneventful readiness period was to end, the QRA section was scrambled to intercept intruding Egyptian Vampires in the area of Sde Boker, the kibbutz in which Ben Gurion was a member. Although it was dusk, the leader of the Israeli pair, Segen David Kishon, on his first operational sortie, observed a pair of Vampires near Ramon Crater. The sun in those hours of the day is on the west and quite low, and Kishon was able to pull up and turn into the Vampires that were probably unaware of their existence. Kishon in Ouragan 323 fired a long burst at one of the Vampires, which caught fire. The second Vampire reversed into the Israeli interceptors, but after a few inconclusive maneuvers withdrew to the west into Sinai. The Vampire on fire crash landed near Abdat, a few kilometers south of Sde Boker, and its pilot, Lt Lotaffi, was taken POW, spending almost a year in Israel before returning to Egypt after the Sinai Campaign.

By the time Segen David Kishon was credited with the first kill ever by a French jet fighter, neither the Ouragan nor the Vampire were the most capable fighters in the inventory of the respective air forces. The EAF had already operated the MiG-15 since late in 1955, while just 24 hours before the above combat, the first six Chel Ha'avir Mystère IVAs landed at Hatzor. In anticipation of the supply of these aircraft Rav Seren Moti Hod succeeded Rav Seren Benjamin Peled as the commander of 113 Squadron, while Peled headed a group of pilots sent to France in March 1956 for conversion to the Mystère IVA with the Armée de l'Air at Cambrai. The unit chosen for converting to the Mystère was the first IDFAF fighter outfit, 101 Squadron, which was disbanded as a Mustang unit at Ramat David. The Mustangs were passed on to 105 Squadron, which then acted as an OTU, also at the same base, while the Spitfires, which until then carried the scorpion badge and the black-yellow fin stripes of this unit, were finally retired from front line service in April 1956.

By this month all the political obstacles were cleared and the first half-dozen Mystères were ready for ferrying to Israel. Each of the $360,000 fighters was to be flown by a young Israeli pilot across the Mediterranean with only a single refueling stop at Brindisi, in southern Italy. The French did not wholly trust the Israelis and offered an Armée de l'Air flight leader as on the Ouragan ferrying, but the confident Benjamin Peled insisted on his leading the flight, being backed by Ezer Weizman, who was responsible for the ferrying operation as the commander of Hatzor air base since February 1956, when he left his previous post as the Ramat David air base CO. The safety margin in the operation was quite narrow as the

For a short time the Ouragans sported the natural metal finish reserved for the advanced fighters. As upon arrival of the Mystère they got their brown and blue camouflage scheme; here a 113 Sqn machine gets a kerosene load from a fuel bowser. (IAF Magazine)

Thus curious picture shows a Mosquito FB.6 with the 119 Squadron insignia on its crew door, when in fact a few of these and other variants served with "The Bats"

Mystères, even with their pair of underwing 620-liter drop tanks, had a reserve for only 15 minutes of flying, and this optimistic result emerged from a calculation with the most favorable conditions, that is, a westerly wind of about 100 knots at the cruising altitude of 30/35,000 ft. Luckily, of the 60 Mystères that were ferried to Israel during the spring and summer of 1956, only two had to divert to RAF Akrotiri in Cyprus, some 400 km northwest of Hatzor, due to critical fuel shortage, and not a single machine was lost.

On 10 April the weather forecast was favorable. Sgan Aluf Paul Kaider, the Air Attaché in the Israeli Embassy in Paris, was responsible for the initial stage, while Aluf Mishne Ezer Weizman took control at Brindisi. A 103 Squadron Noratlas flown by Seren Ori Yaffe served as a communications relay. The six Mystères left Mont de Marsan in the early morning hours of 11 April and flew uneventfully to Brindisi, where they refueled and continued to Hatzor. Peled and the other five pilots were welcomed by the Prime Minister and Minister of, David Ben Gurion; the Minister of Treasury, Levy Eshkol; and the French Embassdor and Chel Ha'Avir's C.O. Aluf Dan Tolkovski.

The arrival of the Mystères fulfilled Tolkovski's requirements of receiving advanced fighters roughly equivalent to the MiG-15 before the early summer of 1956. However, the Mystère was still considered slightly inferior to the MiG-15 and it was feared that the ability of the Soviet fighter to outclimb the French fighter would be critical in combat. Furthermore, a dozen of the Mystères—another six had been ferried in early May—was certainly not an adequate quantity to counter some one hundred MiG-15s. A contract for yet another 12 Mystères was signed on 23 April, and one month later 101 Squadron at Hatzor had 24 Mystères on strength, and it was formally reactivated in May. An intensive period of training commenced with many Ouragan pilots converting to the Mystère while still serving with 113 Squadron, comprised of air-to-air and air-to-ground tactics, using the 30mm DEFA guns and 68mm SNEB rockets.

Gradually, the IDFAF was phasing out of service the veteran piston types. In April 1956 the Spitfire was withdrawn from service and replaced in the OTU role by the Mustang. Simultaneously, the Hatzor-based Mosquito squadron was disbanded, although the PR.16s continued to operate, and moved to Tel Nof where a new Mustang squadron was formed. This Emergency unit was manned by the instructors of the Flying School and its commander was Rav Seren Itschak Evene (Flying School's CO). The days of the Mustang in the OTU role with 105 Squadron were numbered, however, since in September 1956 it was decided to skip the Mustang stage and convert pilots directly from the Harvard to the Meteor. This change did not become fully effective until early 1957 due to the rising tension in the area.

Chel Ha'avir's order of battle for the early summer of 1956 included three jet fighter units with a strength of well over 60 machines, which was a respectable number unless it had to face three, and possibly four, rivaling air arms. The EAF was the most dangerous opponent, while the SAF, IrAF and RJAF presented a lesser threat. Moreover, Egypt and Syria signed an alliance agreement which was undoubtedly aimed against Israel. The Jordanians also had close ties with Egypt, and following the introduction of Soviet equipment, a squadron of Vampire FB.52s was donated to the RJAF as its first front line fighter. A single jet squadron was also on the strength of the SAF, which consisted of Meteors. The IrAF was a larger organization and its jet fighter force included several squadrons equipped with Vampires and Venoms. Obviously, the EAF was the largest and best equipped air force in the Middle East, and in addition to the new MiG-15s and Il-28s, the Vampires and Meteors were still kept in front line service.

A 113 Squadron pilot besides one of the unit's sharkmouthed Ouragans. (GPO)

So Israel had to make yet another step towards the desired numerical balance in spite of the fact that another problem arose—lack of trained jet pilots—and thus training took on a higher pace. The 36 additional Mystère IVAs were procured from France and were ferried to Israel in July 1956 in Operation Geut (High Tide). For political considerations the secret of this new batch had to be kept, and the Italians were informed that severe technical problems in the newly acquired Mystères necessitated the return of 18 of them to the factory in France for repairs! Yet another problem was to assemble a group of 18 experienced Mystère pilots for the ferrying and the benefits of the mass conversion of Ouragan pilots to the Mystères came to good effect for the first time when about 20 pilots (in case any would not be fit) of both squadrons were sent to France. The ferrying of the second 18-aircraft batch caused no problem as they received the same serials as the first group, being flown to Israel as the "repaired" Mystères. Although the Mystères arrived at Hatzor, they were intended for a new squadron at Ramat David. The activation of that unit had to wait until early 1957 due to two factors. The first was corrected even before the Sinai Campaign of October 1956 when the runways at Ramat David were lengthened for the Mystères, which needed more concrete than previous fighters. The second problem was much more critical, and it was a pure lack of jet fighter pilots forced the new Mystères into storage until more jocks would become available.

The procurement of other types involved a small number of each: six Ouragans to augment 113 Squadron at Hatzor and six Meteors for a new squadron to be established at Tel Nof. As might be expected, forming a new unit with such a small number of aircraft implies that these were of a unique nature, and indeed the new Meteors were ex-RAF NF.13 night fighters. Roughly simultaneously, a small number of the same type were supplied to both Egypt and Syria. These Meteors were required against the threat of the EAF Il-28s, which were feared to be able to penetrate Israel's air space at night without any real opposition. In previous years the Mosquito NF.30s saw only limited use and were useless against the fastest Soviet jet bomber. In addition to the procurement of the NF.13s an air radar system was acquired in France, arriving just before the Sinai Campaign. Yehuash "Chato" Tsidon was chosen to form the new unit, and together with a navigator was sent to a night fighting course with the RAF. Another pair was also sent for the same purpose; the second pilot was Ezra Aharon, while the two navigators were Oded Arez and Rachamim Melamed, but they went so near to the Sinai Campaign that they missed it completely. Moreover, only three Meteor NF.13s arrived before and during the period of conflict, and thus it was decided to stop the organization of the new squadron, which surely could not attain operational status before hostilities broke out once more, and a number of Meteors from 117 Squadron at Ramat David, then under the command of Aharon Yoelli, were detached to Tel Nof just prior to the Sinai Campaign to form an emergency unit led by "Chato."

September 1956 came, the tension in the borders was at its peak, and the IDFAF participation in the IDF's retaliation operations of the pre-war period was quite limited as usual. On the night of 10/11 September the Jordanian police station at El Rahwa, south of Hebron, was attacked while a Piper Cub acted as a communications relay. Three nights later it was the turn of the Jordanian police station at Arandal, some 70 km north of Eilat, but this time 100 Squadron fielded no less than five Piper Cubs for the comm relay mission, while yet another five were at readiness for possible evacuation of casualties. The Israeli force did not suffer any casualties against 16 Jordanians killed, and the Pipers were not scrambled. The night of 25/26 September saw a third Israeli retaliation attack;

Two 113 Sqn "The Lions" drive in a scooter to the Ouragan flight line. Actually, the squadron still lives in the attack mission, but flying another type of deadly machine, the McDonnell Douglas AH-64 Apache.(GPO)

The wreckage of the two Egyptian Vampires shot down by Seren Jacob Yoelli on 1 September 1955 in a Meteor FR.9 of 117 Sqn and which were the first jet versus jet kills for the Chel Ha'avir.(IAF Magazine)

the target was Hossan, near Bet Lehem. Two of the faithful Piper Cubs served in their usual relay role while another pair were used for observation along the Bet-Lehen-Daharia road in case any kind of reinforcements appeared. These attacks were performed in direct retaliation to Arab raids in which nine Israeli civilians and seven military had been killed and many others wounded.

The killing of five civilian workers at Sdom on 4 October and of two agricultural workers near Tel Mond, east of Netania, on 9 October produced a retaliation operation in which the Chel Ha'avir fielded, for the first time since the attack on Kuntila, other aircraft than the Piper Cubs. The target was the Jordanian police station in Kalkilia, east of Tel Mond, and it was attacked on the night of 10/11 October. As during previous attacks the Piper Cubs supplied communication relays and observation services. The police station was blown up as usual, but this time the Israeli force that blocked the road from Azun to the police station got messes with superior Jordanian reinforcements and a mobilized unit that was sent to its aid got lost. The serious situation needed a drastic intervention, and the IDFAF was called for help. However, the air force was basically in those days a daytime force and providing CAS at night with the limited aids available was probably either useless or dangerous to the friendly forces. But the seriousness of the situation left them no choice, and a pair of Mustangs from the emergency unit at Tel Nof were scrambled, as well as two Harvards from the Flying School also at the same base. The four aircraft were only able to make low level runs over the area to scare out the Jordanians. Anyway, the restricted aerial support did not help reduce the number of casualties on the Israeli side. When all the forces finally returned to Israel before dawn it was found that 18 Israeli soldiers were killed and 68 wounded. The fact that the Jordanians lost 88 killed and 15 wounded was not a consolation, and the attack on Kalkila was the last retaliation operation—next came the Sinai Campaign.

Just when all the attention was at the Jordanian front, a clandestine diplomatic effort was on its way to create the strange alliance of Israel, France and the United Kingdom against Egypt. The British and the French were the previous owners of the Suez Canal, which had been nationalized by President Nasser. Moreover, the British lost their bases at the Suez Canal Zone and the Egyptian arms market in favor of the Eastern Bloc. The French were also suffering from Nasser's support of the Algerian rebels, and thus the common British and French aspiration was to repossess the Suez Canal and overthrow Nasser in favor of a Western affiliated government. The Israeli interest was to open the Strait of Tiran and destroy the strong Egyptian armed forces, and thus the aspirations of the three countries coincided. As a result, diplomatic talks were held, mainly in France, during late September and early October to define the amount of cooperation among the three allies. The outcome of these talks was that Israel was to attack Egypt in the Sinai by an airborne operation as close as possible to the Suez Canal. The Egyptians would be led to think that this was just another retaliation operation.

However, the British and French, who would mass their forces in the eastern Mediterranean, should claim—as a cover story—that the presence of hostilities so near to the Suez Canal, an international sea lane, risked shipping in the area. Then they would present Israel and Egypt an ultimatum to withdraw from the Suez Canal Zone. Israel would respect it and obviously Egypt would reject it. As a consequence, the British and French forces would destroy the EAF and then land at the northern entrance of the Suez Canal, near Port Said and Port Fuad, and advance south to the Gulf of Suez along the 160 km length of the canal. This was the background to the Suez crisis of late October and early November 1956, as the British and French turned it, or Operation Musketeer as it was known to the military forces. To Israel it has been known ever since as the Sinai Campaign.

Although the British and French plan generally suited the Israelis, its government granted its participation in massive immediate support and Ben Gurion insisted on the presence of Armée de l'Air units in Israel against possible attacks by EAF Il-28s during the first phase of the operation before the Anglo-French intervention. The French agreed, and the massive military supply included a number of Armée de l'Air Dakotas, on loan, to augment 103 Squadron. Other items included ammunitions, fuel and technical ground equipment as the IDFAF had enough quantities for 20/30 days of fighting. The Armée de l'Air units that were to help the Chel Ha'avir in keeping Israel's skies clear of Egyptian bombers were Escadron de Chasse 1/2 "Cicongnes", home-based at Dijon-Longvic, which supplied air and ground crews, using 36 Israeli Mystère IVAs from Ramat David, of course with Israeli markings, being given a bogus designation as "199 Squadron" and "201 Squadron." Also, Escadrons de Chasse 1/1 "Corse" and 3/1 "Argonne" operated F-84F Thunderstreaks from Lod International Airport.

The less than two months warning allowed the Chel Ha'avir to organize for the anticipated second run. Under a thick cover of secrecy, Bedek Aviation was instructed to get ready for action the piston warplanes in storage at Lod which were either to augment existing units or form the basis for new ones. Jacob Ben Chaim, then in staff duties but previously commanding officer of the B-17 unit, was ordered to reactivate 69 Squadron at Ramat David with the three faithful bombers. Also, at the same base Yechezkel Somech, who led the last Mosquito squadron at Hatzor until its disbandment earlier that year, organized a new Mosquito unit, while Moshe Tadmor, commanding officer of the resident 105 Squadron, struggled with Itschak Evene for the available Mustangs. Also at Ramat David was 117 Squadron, with Meteors and with Aharon Yoelli as C.O.

Yehuash Tsidon had been designated as the first CO of 119 Squadron, which with its small number of Meteor NF.13s was working up at Tel Nof. The process was stopped because of the anticipated campaign and the air and ground crews were posted to other units. Another unit at Tel Nof was the Mosquito PR.16 Reconnaissance Flight which was under the command of Elli Ayal. In addition to these, Tel Nof was the home of the IDFAF's only transport outfit, 103 Squadron, which had most of its fleet formed by Dakotas, but also had three Noratlas. It was led by Yehuda Arbel, who was in the U.S. with Ori Yaron, the intended C.O. of the new Helicopter Flight that was to be formed with the remaining Hiller 360 of the two operated by 100 Squadron since 1951, and a pair of newly acquired Sikorsky S-55s. Both officers were sent to the U.S. to convert to the new type, which were to be delivered by sea towards the end of 1956, and to study heliborn operations. Once it was clear that hostilities were imminent they were called back, but arrived after the Sinai Campaign had been launched. Thus, until Arbel returned, Jacob Avishar commanded 103 Squadron.

The Chel Ha'avir's Flying School at Tel Nof was surely not a fighting unit, but yet it formed the basis for three squadrons. Itschak Evene led an emergency Mustang squadron (he was the Flying School CO) that was manned by flying instructors, while Moshe Ashel, the commander of the advanced stage, led a Harvard unit. The third was a liaison squadron intended to support 100 Squadron, especially in communications flying, behind the frontlines. The CO of the Stearman unit was Meir Shefer deploying to Ramle, home of 100 Squadron, who was led by Misha Kener and mainly operated the Piper Cubs. Yet further to the south, Hatzor and its two resident jet fighter squadrons formed the spearhead of the IDFAF combat force. The base was commanded by Ezer Weizman, while 101 Squadron was led by Benjamin Peled and 113 Squadron by Moti Hod. All these officers would have brilliant careers in future years.

On 5 October the IDF's General Staff Operations Department issued the first order concerning the anticipated campaign. The Chel Ha'avir's missions were defined as: destroying the EAF in its bases to achieve total air superiority, the protection of Israel's air space, and the support of ground forces. The IDF plan was to drop the 890th Paratroop Regiment under the command of Rafael "Rafol" Eythan (the IDF CoS during the Lebanon War of 1982), on the western approaches of the Mitla Pass, some 40 km west of the Suez Canal. In these days there were four main roads crossing the Sinai from east to west: the north road ran from the Gaza Strip via El Arish and Romani to Port Said along the coast. The second ran from Nitzana via Abu Ageila, Bir Hama and Bir Gafgafa (Refidim) to Ismailiya. The third also started at Nitzana and went via Qseime and the Gidi Pass to the Bitter Lakes on the Suez Canal. The fourth started at Kuntila and ran via Nahkl and the Mitla Pass to Suez City on the southern mouth of the Suez Canal. The importance of these roads in Sinai, a dry desert, could not be measured, and who controls them rules the peninsula. The two mentioned passes in particular, Mitla and Gidi, prove the point: both were narrow roads, which ran between mountainous ridges, while beyond the sand dunes denied the movement of large mechanized forces. The dropping of a paratroop regiment in that sensitive point was a threat to the Egyptian forces in central Sinai, while the presence of Israeli troops so close to the Suez Canal provided Great Britain and France the desired excuse to present an ultimatum to Egypt and Israel.

The operational plans envisaged that shortly after the 890th Regiment was parachuted, its parent brigade under the command of Ariel Sharon would advance towards the Mitla Pass via Kuntila and Nahkl, while other IDF forces would break into the northern sector of Sinai in the Nitzana-Abu Ageila area, and a lone brigade would advance south along the western coast of the Gulf of Aqaba to Ras Nasrani and Sharm el Sheikh to open the straits of Tiran to Israeli shipping. In the afternoon of 6 October, a Mosquito PR.16 was sent on an operational reconnaissance mission to the Mitla Pass. The photographs obtained clearly illustrated a small camp in its western side, and the interpreters tried to ascertain whether it was a military camp or a party of road workers—if it was the former then the dropping of the paratroopers there was too risky.

During the following fortnight political talks with France and Great Britain resulted in the redefinition of the IDFAF missions. It was agreed that the RAF, Armée de l'Air, FAA and Aéronavale would destroy the EAF in its bases and thus by 21 October it became clear that in spite of Tolkovski's wishes, the Israeli Air Force would not be allowed to attack the EAF. During the first two days of the conflict, before the British and French joined in, Chel Ha'avir's activity would be restricted, in order not to alarm the Egyptians and strengthen the impression that the Mitla Pass drop was just another large scale retaliation operation, and wholly dependent on EAF's activity. By that date only a small number of high ranking officers, less than ten, within the air force knew for sure what the intense activity was all about, while the others were left in ignorance. Nevertheless, on 25 October Avshalon Order (son of King David) was issued by the air staff to the units ordering the immediate commencement of an intense training scheme without mentioning the cause. The tension of the Jordanian front and large retaliation operations of September and early October probably focused the eyes of this area and all accepted the new alert as just a routine issue. The same day the IDF CoS and the Chel Ha'avir CO met, and as a direct consequence a new order was issued to the latter on the night of 25/26 October with a clear definition of its new missions. During the maiden stage of the conflict (the first two days) the aim was to minimize air activity in order to avoid a possible EAF attack on civilian centers. During the second stage, from the third day onwards and after the elimination of the EAF by the British and French, the Chel Ha'avir was to support the IDF ground forces, defend Israeli skies and be ready for the possible intervention of other Arab countries. Besides, it should provide the usual transport and liaison services.

Also, Moshe Dayan granted permission to Yehoshafat Harkabi, head of the IDF Intelligence Department, to approve operational reconnaissance sorties, with many missions being flown just before hostilities broke out. Most of these flights were to Sinai, but one daring long range mission of four hours' duration was flown by Ze'ev Tavor and Alisha Galon in a Mosquito to locate the EAF Il-28 unit. The Beagles were positively identified, with the help of a pair of binoculars, at Luxor, some 700 km from Tel Nof.

Following the issue of the definitive order, the mobilization of the reserves commenced and was totally completed by the IDFAF in only 43 hours, much less than in the other units of the IDF. While this happened, a few dozens of senior leaders gathered on Saturday 27 October at the auditorium of the Chel Ha'avir staff camp at Ramle. Dan Tolkovski himself briefed the astonished officers about Kadesh Order. The secrecy of the operation was still preserved, however, and the commanders were instructed not to reveal the details. Another briefing the same day was at the paratroop brigade in which also the CO of 103 Squadron as well as the senior flight commander and navigator, Eyal Achicar, took part, who was instructed to plan a route to the Mitla Pass. 14 years later Achicar bailed out of his stricken Phantom west of the Suez Canal to be taken POW during the War of Atrittion.

On 28 October another reconnaissance mission over the Mitla Pass was executed, and the photographs revealed that the camp consisted of three huts and 27-29 tents. As a consequence, it was decided not to gamble and the drop zone was moved to the eastern approaches, the less desired but the safest option. At 2000 hrs that evening Dan Tolkovski presented the final Mahkbehs (Press) Order to the same forum that attended the first briefing. It was the code name of the plan to drop the 890th Regiment at the Mitla Pass, which was to include a deception by a Dakota which was to drop dummies near Ismailiya and a pair of Meteors that were to fly at high altitude along the northern coast of Sinai to attract the attention of the Egyptian radar operators. The real effort were the 16 Dakotas, flying as low as possible, escorted by ten Meteors and 12 Ouragans, while 12 Mystères were to orbit above Kabrit to assure that the MiG-15s based there would not reach the Dakotas if scrambled. Once the paratroopers landed, accompanied by Yekotiel Allon as their air support liaison officer, they were to construct a small landing strip from which wounded could be evacuated by a single Dakota and eight Piper Cubs which were on readiness for that mission.

However, three missions were omitted from the widely released order: the cutting of the Egyptian telephone wires in Sinai by a quartet of Mustangs prior to the paradropping, the continuous reconnaissance missions to be flown over the area since the morning of 29 October, and a clandestine mission to eliminate Field Marshal Abed el-Hachin Amar, the CoS of the Egyptian Armed Forces. Amar was visiting Jordan and Syria to seek their support in the face of the Anglo-French threat. He was accompanied by a group of senior staff officers and a few journalists, and the delegation, which was due to leave Damascus back to Cairo on the evening of 28 October, flew in two EAF Il-14s. At 1400 hrs that day Yehuash "Chato" Tsidon, the intended commander of the new 119 Squadron whose activation had been frozen, was ordered to prepare a Meteor

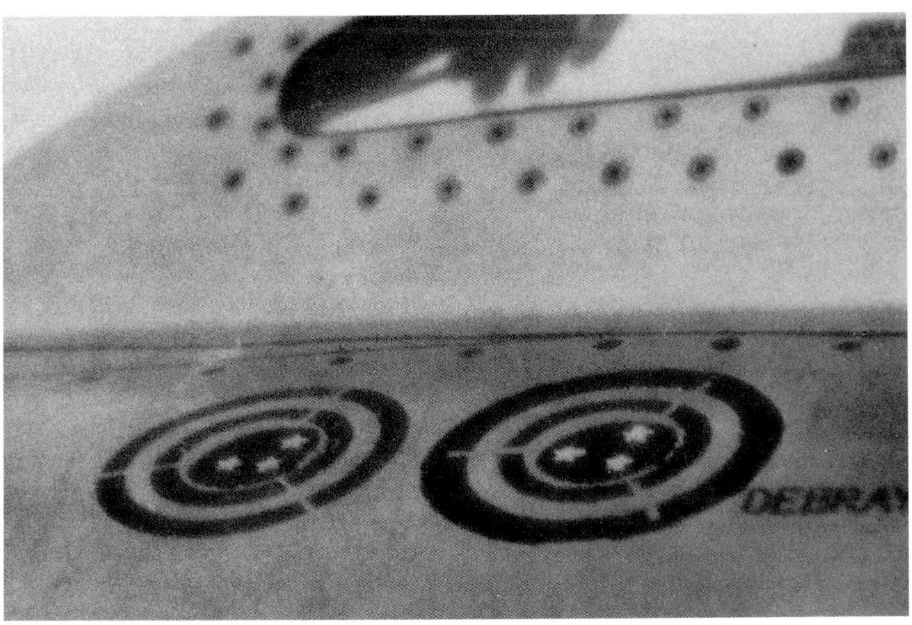

The two kill roundels applied to the Meteor FR.9 flown by Jacob Yoelli.(IAF Magazine)

NF.13 to intercept Amar's Il-14. Chato recognized the problematic execution of that order as during the night interception the fighter had to fly at the same speed as its target for a successful interception, but while the cruising speed of the Il-14 was 175 knots the Meteor NF.13 stalling speed was 140 knots. It implied that aiming would be difficult due to the high angle of attack required to keep the faster jet fighter after the slower transport. Moreover, the fire of the four 20mm guns did slow the Meteor still more, while the surging of hot air from its engines also meant a dangerous reduction of power at a very low speed just above stall.

Nevertheless, "Chato" readied one of the two-seat night fighters and his navigator, Alyasheeb Brosh, practiced that same afternoon with a slow piston aircraft. That last hour of training gave "Chato" the required confidence, and when the lonely Meteor took off Tsidon was determined to fulfill the mission. The night was dark, there was no moon, and Israel was completely blacked out. The Meteor headed west at 10000 ft above the Mediterranean and commenced the search aided by a ground station. The auxiliary fuel system failed—"Chato" could not use the two 300-liter drop tanks, so he jettisoned them and continued his mission. The ground controller directed "Chato", and at a range of three miles Brosh picked the target on his radar screen and directed "Chato", who identified the target visually only at a range of some 200 meters. Dan Tolkovski ordered "Chato" on the radio to obtain a positive identification of the target, so he closed to a range of three or four meters and flew along the slow transport for ten long minutes before he radioed back that the identification was certain. "Chato" opened fire with flaps lowered at a very slow speed and immediately entered a spin as one of the four guns did not fire and the asymmetrical moment caused the Meteor to slow down, stall and enter a spin. "Chato" recovered and saw that the Il-14 was still flying, although its left engine was on fire, which made it go slower, and the Israeli pilot had no choice but to close at higher speed with the hope that the spin, as in the previous gunnery run, would save him from colliding with his target. At a range of 20 meters he opened fire and immediately entered a spin. While going down he noticed that the Il-14, engulfed in flames, was also spinning down. The mission had been completed, although the Meteor was also almost lost as "Chato" recovered only at an altitude of some 300 ft above sea level.

Tsidon's kill was the fourth by a jet fighter in IDFAF service, and it was the first, followed by many more, of the new squadron, nicknamed "Tayeset Ha'atalef" (Bat Squadron), which later flew Vatours, Mirages and Phantoms. It was an amazing achievement although it was marred by the fact that it was not Field-Marshal Amar's Il-14 but the second one which carried 12 officers, two reporters, two photographers and two crewmen. While usually Amar's Il-14 flew first, that evening the order was changed as the official departure ceremony at Damascus' E Maza airport was longer than usual.

By the next afternoon the Israeli Forces would launch in earnest the Sinai Campaign, and hostilities were to flare-up once more in the Middle East. While during the previous conflicts, and the War of Independence the Chel Ha'avir developed from zero, it entered this second round as an organized force which consisted of the following:

TYPE	NUMBER	SQN	AB	NO.PILOTS (1)	CO
Mystère	16	101	Hatzor	21	R/S Peled
Ouragan	22	113	Hatzor	16	R/S Hod
Meteor	15	117	RD	10	R/S Yoelli
Mustang	13	105	RD	14	R/S Tadmor
Mustang	16	116	TN	17	R/S Evens
Mosquito	13	110	RD	16	R/S Somech
Mosquito PR16	3	115	TN	*	R/S Eyal
Harvard	17	140	BS	25	S. Ashel
B-17G	2	69	RD	5**	S/A Ben Chaim
Dakota	16	103	TN		
Noratlas	3	103	TN	31 +13 navigators	R/S Avishar
Piper Cub	17	100	Ramle	24	R/S Kener
Stearman	24	147	Ramle	31	S. Shefer

* - four pilots and five navigators.
** also three navigators and six other aircrewmen.
(Keys= RD: Ramat David; TN: Tel Nof; BS: Be'er Sheva; R/S: Rav Seren; S: Seren.
The number of combat ready aircraft were correct for 29 October; naturally these numbers vary a bit and the following day the CR inventory included, among others: 14 Mystères (out of 37 on strength of 101 Sqn. plus others stored or loaned to EC 1/2 "Cicognes"*), 42 Ouragans and Meteors, 28 Mustangs, 14 Mosquitos, 21 Harvards and two B-17s.
(1) Combat Ready.
*This French unit was named 201 SQN based at Ramat-David and equipped with 12 IDFAF Mystéres. Another French unit was 200 SQN based at Lod with 18 F-84Fs.

The IDFAF top brass included Aluf Dan Tolkovski as Chel Ha'avir's CO; Aluf Mishne Shlomo Lahar, Head of the Air Department; Aluf Mishne Ezer Weizman commander of Hatzor AB; Aluf Mishne Gideon Elrom commander of Ramat David AB and Sgan Aluf Yosseph Ra'annan commander of Tel Nof AB.

EGYPTIAN AIR FORCE ORDER OF BATTLE-OCTOBER 1956

Eastern zone HQ - Ismailia Air Base

5 Sqn, Fayid	12 Meteor F4 and F8
20 Sqn, Kabrit	15 MiG-15
30 Sqn, Abu Sueir	15 MiG-15
31 Sqn, Kasfareet	19 Vampire FB52
40 Sqn, Fayid	12 Vampire FB52
MiF OTU, Kabrit	11 MiG-15UTI
Il-28 OTU, Luxor	21 Il-28

Central zone HQ - Almaza Air Base

1 Sqn, Almaza	16 MiG-15, 6 MiG-17
2 Sqn, Cairo West	16 Vampire FB52, 1 Vampire T55
3 Sqn, Almaza	20 C-47
4 Sqn, Dekheila	light aircraft
7 Sqn, Almaza	20 C-46
8 Sqn, Inchas	12 Il-28
9 Sqn, Inchas	16 Il-28
10 Sqn, Almaza	5 Meteor NF13
11 Sqn, Almaza	19 Il-14

Also six Lancaster bombers (based at Cairo West), as well as eight Fury FB11s and six Spitfire F22s (based at Almaza) were recalled to active duty after having been retired from active service during the previous months.

Those were the forces facing each other. In fact, the important British and French order of battles are missing, but they are not part of the next chapter, which deals exclusively with the role played by the Israeli Air Force in the Suez conflict of 1956, better known as the Sinai Campaign. The IDFAF would leave this second round with the Egyptians as stronger and with more valuable operational experience.

Line-up of Dassault Mystère IVAs for 101 Squadron just after arrival from France, still with the delivery serials. (GPO)

Chapter 9: "Operation Kadesh"

THE SINAI CAMPAIGN OF OCTOBER/NOVEMBER 1956

This was the first modern war in the history of Israel in particular and of the Middle East in general. In contrast to the War of Independence, fought only eight years earlier, it was a war between regular and well organized armies, and the first in which the armored forces played the dominant part. As a consequence, during the eight days of the operation, the IDF ground forces advanced 200 km to the west, very close to the Suez Canal, and about 200 km to the south to capture Sharm el Sheikh and open the Strait of Tiran to Israeli shipping, thus occupying all of the Sinai peninsula.

During Operation Chorev in 1948-1949, the Chel Ha'avir flew 243 operational sorties which was an average of about 15 per day, although this operation had two stages that were separated by a few days of relative inactivity and thus the effective average sorties per day was a bit higher. Even taking into account this fact, however, it was still a very small number compared to the achievements during the Sinai Campaign at an average of 230 per day.

The first day of the campaign, 29 October 1956, was not an indication of the intensive fighting to follow. During the morning hours only a few operational sorties were flown, and these were all reconnaissance missions to the Suez Canal area and the Mitla Pass, while at the air base pilots and aircrews gathered for the briefings. It was a major surprise to the men as secrecy was kept so well that they had no idea concerning the coming war before the initial briefings that were given to 103 Squadron at Tel-Nof at 1000 hrs and to the neighboring Mustang unit, 116 Squadron, at 1115 hrs. Dan Tolkowski took part in both, and although certainly not the spearhead of the IDFAF, both units were to play a key role during the coming hours.

Meanwhile, the reconnaissance pilots were busy checking out the EAF activity at its bases in the Suez Canal Zone, and the paratroopers' drop area at the Mitla Pass, called "Doron" (Gift). Tow reconnaissance sorties were made by Mosquito PR.16s to the Suez Canal Zone, where most of the EAF jet fighters were based. The photographs obtained did not reveal unexpected activity, and all was quiet and normal despite the massive build-up of forces by the British and French and the mobilization of reserves in Israel. The Egyptians were definitively overconfident and unaware of the expected attack. Also, in the morning a pair of 101 Squadron Mystères were dispatched at 0935 hrs for a low altitude visual reconnaissance of the Mitla Pass, which up to then had been surveyed from high altitude. Jacob "Yak" Nevo led Zoric Lev and their report was that the camp at the western approaches was not a military installation. Nevertheless, it was claimed to be too late to change the drop zone, although seven hours were still available between the fighter pilots' report and the planned take-off of the 16 Dakotas. As the noon hours approached, activity grew a little with a third reconnaissance mission by a Mosquito to the Suez Canal dispatched at 1415 hrs from Tel Nof, just fifteen minutes after the first pair of four cable cutter Mustangs from 116 Squadron left the same air base.

Dassault Ouragan of 113 Squadron with the furious looking sharkmouth markings. Note taped over serial and squadron badge on tail. (IAF Magazine)

Number 100 Squadron Piper Cub with the Sinain Campaign identification stripes in its typical environement at Sinai on 2 November 1956.(GPO)

During the mid-fifties, electronic warfare was a fledging science for the Israelis, and the only possible way to deny the Egyptians from their main communication channels was simply to cut the wires. Wireless communications were not in widespread use as in the present, and most armies in the Middle East still relied heavily on the ordinary telephone lines. If these cables would be disconnected the Egyptian high command would not receive the correct battle scene from the local commanders at the Sinai and thus might fail to understand the seriousness of the situation. If that aim could be achieved the deployment of the EAF and the infusion of reinforcements into Sinai would be delayed and allow the IDF ground forces to tackle at first only with the garrison forces. Moreover, the belief that the activity at Sinai was a mere large scale retaliation would cause the Egyptian high command to avoid the use of its Il-28 bombers against Israeli civilian centers in order not to escalate the "local skirmish", and it was helped that by the time the Egyptian command would realize the fact that it was an all-out war the EAF Il-28s would be wrecked by the Anglo-French air raid that was supposed to begin at 0600 hrs on 31 October.

The first proposal was to send a sapper in a Piper Cub that would land near to a telephone line, blow the pole, return to the waiting Piper Cub, and fly back to Israel. This proposal was rejected in favor of a more risky idea: to cut the cables with a Mustang, a mission it was never designed to fulfill, and for which a steel cable 50 meters long and equipped with weights to keep it lower than the aircraft was attached to the veteran fighter. To test the idea a duplication of the Egyptian telephone lines was constructed near Hatzor. The line was composed of 17 cables in four groups of four each and a single cable on top. Amitai Chasson, nicknamed "Shafan" (Rabbit), a Flying School instructor in emergency posting to 116 Squadron, was chosen for the trials. Shafan was to fly at an altitude of 60 ft above the ground, level, with the cable dragging on the ground. Several tests confirmed that the idea was feasible, but that the reliability of the installation allowed only one certain pass.

Once the initial trials were completed Shafan was joined by five other Mustang pilots that practiced on the dummy telephone line.

On Thursday, 25 October, all leaves were canceled and the pilots were ordered to stay at their air bases. Two days later the purpose of their training was finally unveiled to the senior Mustang

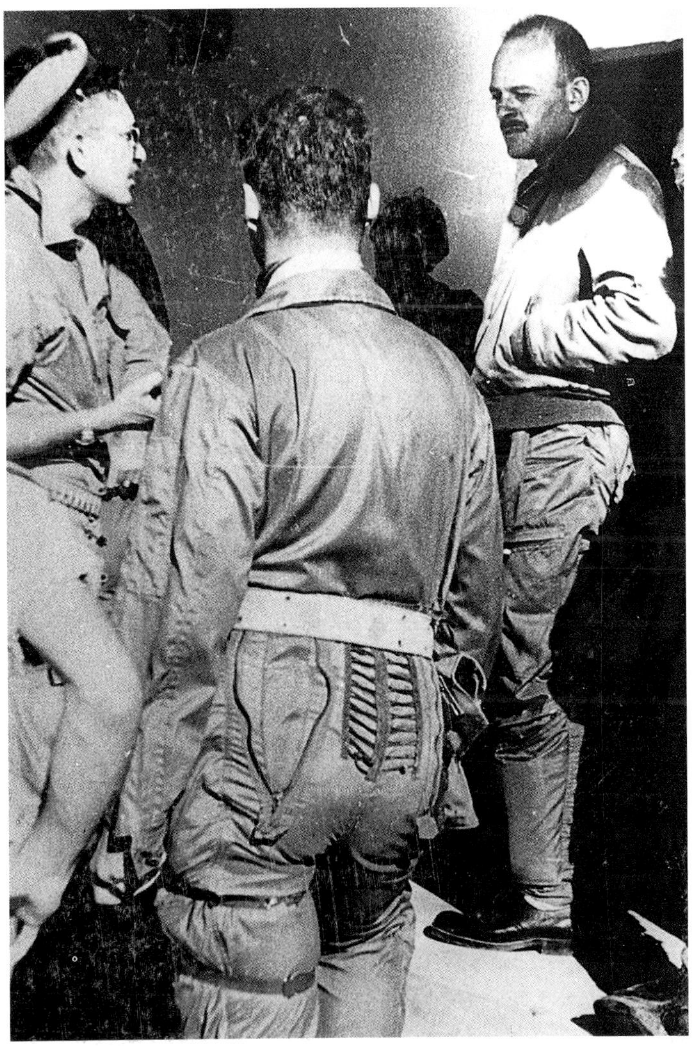

101 Squadron pilots at the squadron's operations room at Hatzor, prior to a mission during the Sinai Campaign. (GPO)

C-47 Dakotas getting ready for the airborne operation over the Mitla Pass in Sinai at the outbreak of the Sinai Campaig. (GPO)

pilots during the briefing at Ramle when Dan Tolkoswki informed the officers that two hours before the 890th Regiment parachuted at the Mitla Pass, two pairs of Mustangs would cut the main Egyptian telephone lines at Sinai. Only on the morning of Monday, 29 October, did the remaining pilots of 116 Squadron at Tel-Nof learn of the nature of the strange activity they witnessed a few days before. At 1115 hrs that day the squadron commander Itschak Evene commenced the briefing of his men while Dan Tolkowski listened carefully.

The four pilots that even picked up, himself included, were the squadron vice-commander Dan "Harry" Barak and two of his flight commanders, Aryeh "Cheattie" Chaitlin and Amitai Chasson. The first section to take-off at 1400 consisted of Barak and Chaitlin, as their target was more distant than that of the other pair, and Evene and Chasson took off minutes later to preserve a similar timing. The target of the first section was the telephone line between Kuntila, the Mitla Pass, and Suez City. The two cutting points of the second section were about 100 km to the east, west of Thamad. The Mustangs flew at very low altitude to avoid detection and in radio silence with full internal ammunition for the six 1270mm Brownings. While en route already deep into Sinai, Chaitlin's steel cable detached, but he nevertheless continued to fly in loose formation with Barak, who entered the planned low run on the telephone line in which his cable also was detached. Nevertheless, Barak was not sure that the telephone lines were torn, so he came in once more at very low altitude and cut the cables with his propeller. It took Barak several runs to overcome the natural instinct to pull up just in front of the cables. Next came Chaitlin's turn in the second point and he made six runs to cut the telephone lines.

Simultaneously, the other section had the same problem and improvised similar methods. The cable from Evene was detached from his Mustang during the first run before he reached the telephone line, so Chasson came in and also lost his cable without completing the mission. Evene also chose to cut the lines with his propeller, and also faced the instinct to pull up just in front of the cables, but finally, after several passes, the cables were cut and they cleared the area to the second location where Chasson's turn came, who disliked the idea of messing up cables in his propeller and tore them with his wingtip. In spite of the difficulties involved, the mission was successfully completed. On their way back to Tel-Nof, the four Mustang pilots saw the Dakotas on their way to the Mitla Pass, the true start of the Sinai Campaign.

More than an hour before Evene commenced the briefing of his pilots, Jacob Avishar commenced the briefing of the Dakota crews at Tel-Nof. 16 Dakotas were to fly at a very low altitude of 100 ft above ground level in four formations of four each carrying the 395 paratroopers. The leader was naturally Avishar, while his navigator was Eyal Ahicar and the co-pilot was Yael Rom, one of

Pilots of 101 Squadron and their Mystères at Hatzor air base. (IAF Magazine)

the two female pilots in Chel Ha'avir service during the Sinai Campaign. The other female pilot, Rina Levinson, was flying Boeing Stearmans on liaison duties. One of the pilots in the formation was none other than Segen (reserve) Pessach Tal, the veteran Dakota pilot who, like many others, left the IDFAF for civilian life. For him it was kibbutz Kfar Giladdi on the Lebanese border, but he continued to fly as a reservist and was mobilized several days before the Sinai Campaign. Avishar stressed to the crews that the mission should be fulfilled, and if the formation dispersed for any reason, be it enemy interceptors or technical difficulties, each Dakota should press on individually and drop the paratroopers in the planned area east of the Mitla Pass. After he finished, it was the turn of the squadron Intelligence Officer (IO) and then the Operations Officer, who read the list of crews and the planned schedule. Next arrived Dan Tolkowski who started a joint briefing with the escorting fighter pilots. The last briefing, besides the Dakotas, was to the paratroopers as well as to the aircrews, and was given by Avishar and Eythan. At 1520 the wheels of the first Dakota, flown by Avishar and with Eythan on board, left the concrete runway and immediately headed south at low altitude with the other Dakotas gradually joining in. The only voice to be heard on the radio was that of Yael Rom, which reported the passage of the check points by numbers "1", "2", "3", and so on. As the last heavily-laden Dakota took off from Tel Nof the deception effort commenced, although the parachuting of dummies near Ismailiya by a Dakota was canceled due to the high risk involved and the general inactivity in the Suez Canal Zone as was clearly evident from the photographs gathered by the Mosquitos the same morning. The pair of unarmed Meteor T.7s that were to fly along the northern shores of Sinai at high altitude to attract the attention of the Egyptian radar operators launched from Tel-Nof at 1533, and, although the leader, Elli Eyal, returned shortly afterwards due to a technical fault, his wingman continued alone with the risky mission as an unarmed decoy. Strangely enough, the lonely Meteor did not attract any attention.

Dan Tolkowski was the Commanding Officer of the Chel Ha'avir during the Sinai Campaign. This picture taken approximately one year before shows him after his last flight in a Spitfire of 105 Squadron, before the unit re-equipped with P-51D Mustangs.

Loading a Meteor F.8 with a rocket load prior to a close air support mission. (GPO)

A crash landed Mustang after being hit by Egyptian flak, these obsolete propeller driven fighters suffered heavy losses during the war, and shortly afterwards with the intake of more jets, were phased out of service.

While the Dakotas were flying slowly south, the parent unit of the 890th Regiment, the 202nd Paratroop Brigade, commenced its advance to Kuntill, and at 1545 hrs the first light-plane to take an active part in the conflict took off from Eilat. Moshe Aven was the pilot of the Piper Cub, and his mission was to support the advancing paratroopers and then continue to the Mitla Pass to support the 890th Regiment. At 1600 hrs two Mystères overflew the Doron area to make sure that all was clear; the leader was Joe Allon, and while he was passing over the Mitla Pass, Aharon Yoelli led the Meteors which were to escort the Dakotas from the Israeli-Egyptian border to the Mitla Pass. The much faster Meteors stepped up between 5,000/10,000 ft, waving to keep a watful eye on the helpless transports down below. The crossing point of the border was west of Ramon Crater, and it was accomplished at 1630, barely five minutes before six 113 Squadron Ouragans launched from Hatzor. Moti Hod was the leader of that unique formation, and three of its pilots would be future IDFAF commanders: Hod himself, David Ivri and Amos Lapidot. Another pilot in that flight was David Kishon, who had been credited with the first Ouragan kill. The 113 Squadron mission was to relieve the Meteors and cover the drop phase itself, which took a while as each Dakota had to make several runs. Simultaneously, twelve Mystères from 101 Squadron also launched from Hatzor for CAPing over Kabrit, while no Mystères were left on QRA against possible EAF retaliation because of the presence of the 2e Escadre de Chasse pilots flying Israeli Mistères as "201" Squadron at Ramat-David. This mission of such a large formation, led by Benjamin Peled, entailed flying at high altitude over Kabrit and engaging any MiG-15 that would take-off. The pilots were ordered not to attack them on the ground, and while circling above the EAF air base they observed that the Egyptians were dispersing out the MiGs that previously were closely parked. No other unusual activity was noticed, and no MiG-15s launched from Kabrit during that crucial dusk.

The dropping started as planned, but due to a navigation error, the paratroopers were parachuted four Kilometers east of the Doron Area and had to walk for about an hour to reach their planned destination. After that was completed, another six Ouragans arrived to relieve the first formation, but owing to the falling darkness the pilots were unable to identify the Dakotas, which turned east and returned to Tel-Nof, leaving behind a radio relay Dakota. The first phase of the Sinai Campaign was accomplished, with a sizable Israeli force firmly established some 40 Kilometers east of the Suez Canal. The British and French got their desired excuse to launch Operation Musketeer. Israel had fulfilled its obligation.

A P-51D Mustang from 105 Squadron gets a rocket load for a mission in support of Israeli troops in the Sinai.(IAF Magazine)

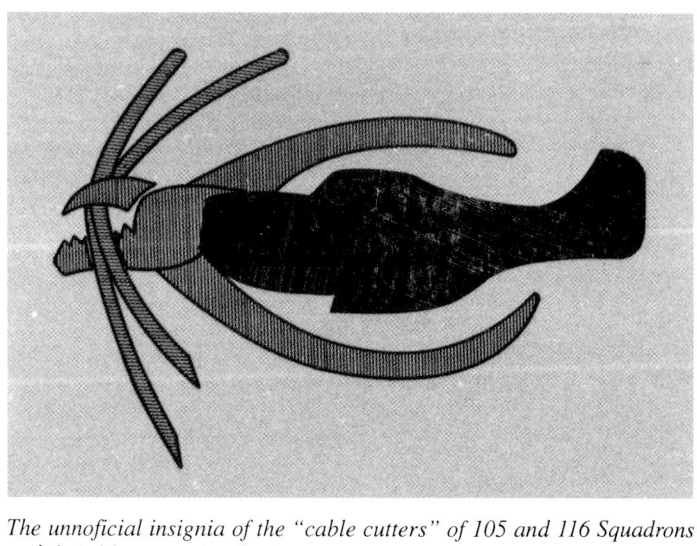

The unnoficial insignia of the "cable cutters" of 105 and 116 Squadrons and their Mustangs. (author)

The victorious pilots of 113 Squadron at Hatzor in from of a camouflaged Ouragan (number 29) with the captured Egyptian flag of the destroyer they had attacked and damaged some days before.(IAF Magazine)

After salvaging and a thorough restoration, this MiG-15 was put on display at Hatzor air base, home of 101 Squadron. Unfortunately, its victor, Yak Nevo, died in September 1989 at the age of 57 from a heart attack.(IAF Magazine.

During the night of 29/30 October the focus was turned on the 890th Regiment spending its first night almost 200 Kilometers deep in enemy territory. The first Piper Cub to arrive to the Doron Area with Rishin Aven at the controls flew from Kuntila and was ordered to survey the roads in the area before returning to Eilat. Yet another Piper Cub was dispatched from Eilat to the Mitla Pass to evacuate a wounded paratrooper, but its pilot became lost and landed in the desert to spend the night alone before continuing the next morning to his destination. All communications with the 890th Regiment were through Yakotiel Allon, a pilot with 100 Squadron, who parachuted with the paratroopers as their regimental aerial support liaison officer.

This scene shows the salvaging operation from the shallow waters of Bardavil, west of El Arish in December 1956 of the Egyptian Air force MiG-15bis shot down by Seren Yak Nevo of 101 Sqn on 31 October; note little apparent damage to the Fagot.(IAF Magazine)

The three Noratlas of 103 Squadron joined the Dakotas from the same unit during darkness, and until 0,00 hrs, together with three Dakotas, dropped additional paratroopers, supplies and 120mm mortars at Doron. Once that effort ended, Armée de l'Air Noratlas became the first Anglo-French aircraft to take an active part in the Sinai Campaign. The French transport, loaded with arms and supplies, took off from British air bases at Cyprus and landed at Tel-Nof, where Israeli navigators joined the French crews. In spite of the fact that two of the Noratlas were unable to locate the paratroopers, the others parachuted eight jeeps, four 106mm recoilless guns and another pair of 120mm mortars.

At dawn on Tuesday, 30 October, the first three Piper Cubs landed at Doron, including Amnon Lolav, the pilot that spent the night in the desert, but his Piper was damaged in a hard landing at the Mitla Pass. While the light-planes were flying at ground level, high above, 101 Squadron Mystères supplied almost constant CAPs over Doron to cover the forces there, which did not have AA guns at their disposal to counter any EAF attack. Naturally, war squeezes the most out of the men, and by Tuesday's dawn, when the first pair of Mystères were dispatched to the CAP station over Doron, the diary of 101 Squadron claimed that the number of serviceable Mystères was twice that of the previous day. The Mystères took off in pairs every forty minutes to keep their CAP, and in spite of the fact that the paratroopers, already in contact with Egyptian forces and under mortar fire, requested close-air support (CAS) , the Mystères stayed at high altitude as ordered. The aim was still to keep the Egyptians under the impression that it was only a small scale operation. Each pair of Mystères stayed over Doran for about ten minutes before reaching bingo fuel, and as the flight from Hatzor took about twenty minutes, the forty minute intervals between the scheduled take-off of the different sections left the paratroopers without cover for about 30 minutes at each interval.

In spite of the cable cutting effort the previous day, the Egyptian local HQ at Nakhl reported back about the passage of Dakotas and the first sign that the EAF was finally waking up was a visual reconnaissance by four Vampires over the Suez City-Kuntila road passing on their way over the 890th Regiment at the Mitla Pass and

The old Flying Fortresses of the reactivated 69 Squadron performed several combat missions during the Sinai Campaign. No losses were suffered, but one of them suffered a direct hit in one wing from another left over weapon from WWII, an Egyptian manned 88mm German flak gun!.(IAF Magazine)

the 202nd Brigade, which was already at Thamad, more than 60 kilometers deep into Sinai but still about 120 Kilometers east of the Mitla Pass. The Vampires were not intercepted. Their report was quickly used, and immediately afterwards two pairs of MiG-15s attacked both Israeli units, causing some damage. At the Mitla Pass, Moshe Bokee was ready to take-off in his Cub with a wounded paratrooper strapped to the stretcher aft of the pilot. The MiG-15s observed the Piper and dived at it, setting the helpless light-plane on fire and wounding the regimental doctor, Samal Agmon. Boker, who escaped from his burning aircraft, returned to it and rescued the trapped soldier. Meanwhile, the MiG-15s made several more strafing runs in which another two paratroopers were wounded. The Mystère CAP did not get a Tally Ho on the MiGs.

Another formation of MiG-15s tried to attack the 202nd Brigade near Thamed at 1000 hrs. The brigade was enjoying the services of a lonely Piper Cub patrolling ahead and reporting back any news. Its pilot was Benjamin Cahana, who received his PPL back in April 1939 as one of the first six graduates of Palestine Flying Services and had served with the IDFAF as a light-plane pilot since the War of Independence. The MiGs observed the Cub and attacked it, but Cahana, instead of escaping, turned into the jet fighters and drew them into an unequal dogfight in order to save the ground forces below from the expected air attack. He probably hoped that the MiG-15s, short on fuel, would clear the area after a short combat against him. Cahana achieved his goal and drew the MiG-15s' attention, but was killed when a burst of cannon fire finally struck the Piper, which exploded

At 1030 the Chel Ha'avir was allowed to give CAS to the ground forces, and the pair of Mystères on CAP over the Mitla Pass, led by Joe Allon with Ohad Shadmi as his wingman, were immediately ordered to attack an Egyptian column on its way to the Mitla Pass area. However, the first such missions were flown during the early noon hours and only by the jet fighters as the EAF was much in presence, and although the Mustang/Harvard/Mos-

A pair of Dakotas seen at a forward airstrip in the Sinai during the war. (GPO)

A Piper PA-11 Cub of 100 Squadron landing "somewhere" in the Sinai desert in support of ground forces.

quito squadrons received CAS orders, they were not cleared to take-off yet. At Tel-Nof, for example, twelve Mustang pilots sat in the cockpits for half an hour before their mission was finally canceled. An operational sortie by piston-engined fighters was nevertheless accomplished using the same dangerous methods of the previous day. Thus, with the Mystère committed to CAPs and the piston fighters grounded due to the fear from EAF jet fighters, the main burden that day fell into the Ouragan and Meteor squadrons. At 1325 hrs eight Ouragans left Hatzor in two formations. The first was sent for CAS to the Mitla Pass, but although the paratroopers claimed to have heard them, the Ouragans could not identify their positions and returned to Hatzor. The other formation was more successful and attacked some 20 vehicles and a camp at Bir Hasana, almost 80 km west of Nitzana, where the main IDF effort to break into northern Sinai was launched on the first interdiction mission of the Sinai Campaign. Later on, until 1700 hrs, which was about half an hour before darkness fell, another three Ouragan formations were dispatched, all to the Mitla Pass. The last, led by Allon, attacked an Egyptian 120mm mortar battery with gunfire and rockets , which bothered the paratroopers.

The Meteors also took part in this activity and were the first IDFAF fighters to engage in combat with EAF MiG-15s when a pair of 117 Squadron machines were on an interdiction mission to Gabel Libni, some 60 km west of Nitzana, being intercepted by five (a strange number for a formation, but this was what was reported by the Israeli pilots) MiG-15s. One of the Meteor pilots escaped immediately, but the other, Hilel Elroee, stayed with the MiGs. After a few maneuvers he tried to release his drop tanks, but only one could be jettisoned, and that technical fault saved him as the Meteor immediately entered an unexpected spin and the MiGs lost contact. Elroee recovered successfully at low altitude and returned to his base. Also, the Meteors claimed another "first" when one of them was hit by Egyptian flak and the victim was none other than the squadron CO Aharon Yoelli, who was attacking vehicles near El Arish. The Meteor was seriously damaged, but Yoelli, slightly injured from fragments, flew back to Tel-Nof on one engine only to find that the right mainwheel could not be lowered., so he retracted the other wheels and performed a perfect belly landing, after which he was immediately taken to the nearby Caplan Hospital. The next day he was back in action.

Finally, in the late afternoon hours, the first true mass aerial combat in the history of the Middle East wars in which jet fighters were involved took place over Kabrit. The Intelligence Department claimed that a massive air attack on the paratroopers at the Mitla Pass was planned by the EAF for the late afternoon hours. The Operations Department immediately ordered to scramble all available Mystères, but Benjamin Peled sent only six to preserve adequate reserves for future developments. The first formation of three Mystères was led by Jacob "Yak" Nevo, with Shai Agozi and Yosspeh Tsuk, who was nicknamed "Falach" (Farmer) and was one of the less experienced pilots with only twelve hours on the type. As they approached Kabrit at 20,000 ft they noticed six MiG-15s taking off and immediately dived at them, opening fire almost simultaneously. Nevo and Shadmi missed, but Tsuk, who was the only one to ease the throttle back and deploy the air brakes to slow his fighter during the dive, fired two long bursts at his target and saw the MiG as it caught fire and dived into the ground, to become the Mystère's first confirmed kill in IDFAF service. But the combat did not end at that stage, as the other three Mystères and a few additional MiGs joined in. It was thought that while the climb rate of the MiG-15 was superior to that of the Mystère, its turning capabilities were slightly inferior to that of the French fighter. Quickly

An Egyptian convoy being strafed by an Ouragan of 113 Squadron. (IAF Magazine)

A Mystère IVA section from 101 Squadron taking-off for an anti-MiG patrol during the 1956 conflict. Note the Sinai Campaign stripes. (IAF Magazine)

the Israeli pilots realized that in practice the MiGs turned as fast as their Mystères did, so the dogfight was inconclusive, and although Tsuk received a 37mm Armor Piercing shell in his right wing, all the Mystères regained Hatzor.

Another evidence from this combat was the identification problem as the MiG-15 and Mystère IVA were similar fighters and the pilots had to report constantly on their positions to avoid mistakes, and while the Meteors—operated by both sides—were exactly identical and the method used during the War of Independence when both antagonists flew Spitfires, that all other formations in the sky must be of the enemy, as 101 Squadron usually mounted only one CAP at a time, was not valid in the more crowded skies of the Sinai Campaign.

Roughly at the same time that the Mystères were fighting with the MiG-15s over Kabrit, a Meteor T.7, equipped with cameras and escorted by a Meteor F.8, was sent for a reconnaissance mission over the Mitla Pass. Meanwhile, five 113 Sqn. Ouragans were busily attacking Egyptian mortar positions opposing the Israelis, but no fighter pilot can resist the temptation of air combat, and as the Ouragans noticed the pair of aircraft high above, they checked with the controller if any other friendly aircraft were in the area. The surprising answer, possibly because the controller was not aware of the reconnaissance flight, was negative, so the Ouragan pilots forgot all about their CAS mission and zoomed up for the kill. Fortunately enough the IDFAF was quite a small organization at the time and the Meteor pilots clearly identified the voices of the Ouragan pilots, so they called them in their private names to advise about the controller's mistake.

As the first full day of fighting ended, the assumption that the EAF would confine its forces to the Sinai peninsula proved correct, and although the IDF forces suffered from EAF air attacks, the number of casualties was small and their advance was not stopped by them. The EAF activity was somewhat limited with only about 40 operational sorties, in which the MiGs claimed two Israeli Piper Cubs, one on the ground and the other in the air, while losing one of their number to Segen Tsuk's Mystère from 101 Squadron over Kabrit.

The night of Tuesday/Wednesday, 30/31 October, was just like the previous night, dominated by the transport crews and the light-plane pilots. The transports flew in singles to drop supplies at the Mitla Pass, but several of the Dakotas were unable to locate Doron because of the layer of stratus cumulus clouds at about 3,000 ft and they could not go lower because of the risk of colliding with the mountains in the area, which were almost 3,000 ft high. Simultaneously, the Piper Cubs were sent to evacuate wounded paratroopers under the cover of darkness.

But the same night also saw an activity of a totally different nature. Up to that stage the piston-engined fighters and bombers had been kept on the ground in anticipation of the Anglo-French attack on the EAF air bases that would eliminate the MiG threat. In fact, their only operational sorties until then were the telephone cable cutting effort by the Mustangs. During the evening hours of 30 October the Mosquitos also joined the operational club, since the EAF night fighting capability was a handful of Meteor NF.13s defending Cairo. Two Mosquitos were dispatched separately to attack vehicles in northern Sinai. The column was located and attacked, but the efficiency of that action can be highly suspected due to the primitive methods employed.

But the Mosquitos were not the only nocturnal predators; four EAF Il-28s were sent to bomb Israeli air bases, but none found their target and at least one bombed open ground near Jerusalem. One of the Il-28s flown by Capt Mostafa Chilmi did not return to its base, although that loss must have been caused by an accident or some kind of failure since it was not caused by IDFAF action—the Meteor NF.13s were not flown operationally during the Sinai Campaign. But early on that morning one of the most bizarre chapters of this war commenced to develop. At 0330 on 31 October the Egyptian Navy destroyer Ibraim el Awal started to shell Haifa harbor with some 220 4-inch rounds. Amazingly, no damage was caused and the Israeli Navy destroyers Yaffo and Eilat were ordered to intercept the Egyptian warship, and at 0527 the first rounds were exchanged. Also at dawn a lonely Dakota was sent to locate the area of the naval combat and direct IDFAF fighters to the scene to support the Israeli destroyers. At 0546 the Dakota confirmed positive identification and a pair of Ouragans were scrambled from Hatzor at 0630 with full internal ammunition and 16 AP rockets each. Jacob Agasi led David Kishon, and they attacked one after the other, each firing his full load of rockets in a single pass while firing the guns to keep the heads down. The effect was apparent; the steering gear was damaged, the electricity supply was lost, and the ammunition elevator went out of action. Ibrahim el Awal's captain requested fighter cover, but not a single EAF aircraft was sent to his aid, and at 0710 a white flag was seen on its mast. Yaffo and Eilat stopped firing and closed on the crippled Egyptian destroyer, taking its crew POW and tugging the ship to Haifa harbor where it was repaired and entered service with the Israeli Navy as Haifa. At 0730 a second pair of Ouragans were scrambled to attack the warship, but while on their way the pilots, Moti Hod and Ezer Weizman, were informed that their target had surrendered and thus they changed to the usual interdiction, attacking vehicles near El Arish and 113 Squadron CO, and Hatzor air base CO claimed six of them destroyed.

Meanwhile, the neighboring Mystères assumed once more their task of CAPing over Doron. The first section dispatched to the area took off before dawn and included Shai Agozi and his wingman Aharon "Yalo" Sahvit. While at 36,000 ft Yakotiel Allon called for help as EAF Vampires were attacking the paratroopers, and the pair dived on the four Vampires. At 15,000 ft the Israeli pilots tally-hoed their opponents flying in an odd vic formation, and Shavit was the first to fire at the rearmost Vampire, which caught fire immediately. Meanwhile the leader, Agozi, gave the same treatment to a second Vampire, and during the short and decisive combat which followed the remaining two Vampires were also downed, each 101 Sqn. claiming one. The two pilots reached bingo fuel and both made successful emergency landings at Hatzor.

The events on the morning of Wednesday 31 October suited the Chel Ha'avir's wishes admirably; four kills and the capture of a destroyer were great achievements. However, the British and French delayed their joint attack on the EAF and thus, instead of flying in a MiG-free sky, it had to face the EAF alone for one more day. Yet another consequence was that the piston aircraft had to operate in spite of the MiG threat to support the IDF ground forces which were pushing in northern Sinai. The heavy burden lay especially on the two Mustang squadrons that operated around the clock against heavily defended targets, and if the Mustangs proved to be highly vulnerable to flak due to their liquid cooled powerplant, the Harvard proved even worse.

At 0800 the Harvard's first operational mission in the Sinai Campaign was launched when four of them, led by Moshe Ashel, left Be'er Sheva to attack Egyptian strongpoints in Om Kattef, south of Nitzana. All four Harvards were damaged by flak, and only two regained their base. The third made an emergency landing in a field, while the fourth crashed with the loss of the squadron CO, Seren Ashel. It was a major blow to 140 Squadron, an improvised emergency unit, and the second formation which was sent to the area fared no better, although neither of its aircraft was damaged as the target was not identified. As a result of the loss of its CO, activity in the squadron was not resumed until noon, when Seren Ovadiah Nachman was appointed as the new CO.

While Harvards were attacking Om Kettef the first formation of Mustangs were also in action for the first time in the CAS and interdiction roles. The first formation of 116 Squadron attacked Egyptian armored forces near Bir Gafgafa, almost mid-way between the Israeli border and the Suez Canal, losing one of them to flak. Its pilot, Elded Paz, climbed to 1,300 ft but decided to land instead of bailing out, possibly because of the low altitude involved. Later he said that he was close to the Egyptian forces and he feared that they would shoot at him while descending in the parachute. He crash landed successfully and immediately started walking east, meeting Israeli troops near Jabel Libni after more than 60 kilometers covered during 37 hours of march.

The same unit lost another Mustang to flak during the morning when Ori Schlesinger received small arms hits in his engine. He flew for six minutes to gain Israeli controlled terrain, but was killed during an unsuccessful crash landing near Chalutsa in the northern Negev. In spite of these losses, the value of the piston-engined fighters effort was important. The main weapons used during the CAS and interdiction missions were rockets and napalm bombs.

Going back to jet fighters, 113 Squadron had launched since the early morning nine Ouragan missions amounting to twenty sorties, and it was to be the tenth mission, launched at 1245 hrs, that would attack Egyptian forces nearby in the Mitla Pass, that met EAF MiGs in the air, although 101 Squadron Mystères had already met them twice earlier that day. Yak Nevo led Yossepth Tsuk on a CAP at the Mitla Pass and on their return, roughly between 1000 and 1100, they engaged in combat with seven MiG-15s, of which four left the area, leaving three to counter the two Israeli jets. Yak turned with the three MiGs for some ten minutes while Tsuk was keeping a close eye from above. The combat was fruitless, how-

For most of their missions, the Mystère IVAs of 101 Squadron flew Combat Air Patrol sorties over the Sinai Campaign, providing cover to Israeli ground forces. (IAF Magazine)

Late in their career the B-17G Flying Fortresses of 69 Squadron saw their chin turrets replaced by a search radar, and in this guise added another role to their tired airframes, that of maritime patrol and surveillance.(IAF Magazine)

ever, and although Yak fired a few bursts from his only operative 30mm cannon (the other had jammed), he scored no hits and was frustrated to observe that one of the MiGs recovered from a right hand spin. The Israeli pilots were led to think that the MiG-15 could not recover from such a spin!. On their way back the pair was joined by another two Mystères that flew home after expending all their ammunition in a CAS to the paratroopers at the Mitla Pass. All four were bounced by MiG-15s from six o'clock but escaped the first surprising pass undamaged, and Yak was the only Mystère pilot to chase a MiG. He opened fire from 400 meters and missed, so he closed to 200 and this time he scored hits, damaging the MiG who lost some pieces and started to pour white vapor, indicating a fuel leak. To finish it off Yak closed to 50 meters and squeezed the trigger to find that he was out of ammunition, as were 11 other Mystères in the area. The damaged MiG headed to El Arish air base, but Yak, followed by Zoric Lev, who was the leader of the other pair, crossed his flight path while he was on finals and the frightened Egyptian pilot abandoned his landing attempt and continued west while losing height. The Israeli pilots had to return to Hatzor, and only after the Sinai Campaign ended, that particular MiG-15 was found at Bardavil, northern Sinai, almost intact in shallow water, and was brought to Hatzo where it is still preserved.

Almost three hours later Seren Nevo was back in action. A pair of Mystères got messed in a tough combat with a few MiG-15s, and although one MiG was damaged in its tail, no kills were scored initially in that combat as six Mystères were sent to help the fighting pair. Among the pilots was again Nevo, who as usual had Tsuk as his wingman. They observed two MiGs and made a bounce on one of them with Tsuk the first to achieve advantage, but he was unable to fire as the angle-off was too much, so Yak came in and opened fire from 200 meters. The MiG entered a spin at 18,000 ft, from which it recovered 8,000 ft lower, when its pilot ejected and the burning fighter dived to the ground. It was Yak's second kill that day. While the Mystères enjoyed a relatively easy kill, mainly attributed to the fact that it was Yak's fourth air combat in less than two days, thus gaining experience after each one, a pair of Ouragans were fighting hard for their lives. At 1245 a pair of Ouragans launched from Hatzor on an interdiction mission to the Mitla Pass area. However, due to malfunctioning fuel systems which rendered the wingtip tanks unusable, the mission was changed to interdiction on the road between Bir Hasana and Jabel Libni. Just before attacking they observed several swept wing fighters, but thought they were friendly Mystères. The assumption proved wrong, and the second Ouragan was immediately bounced from its rear quarter. The pilot jettisoned his load, napalm bombs, rockets and wingtip fuel tanks, and broke hard, counting six to eight MiGs on his tail. He warned the leader, who was critical on fuel due to the abovementioned malfunction, and thus while his wingman was messing up with the MiGs he escaped home at low altitude, although he did not make it to Hatzor. Near Chalotsa the engine stopped, and he made a perfect dead stick belly landing on the sandy floor. The wingman called for help, and a pair of Mystères on CAP were sent to the area and the lonely Ouragan was at least able to escape at low altitude along Jabel Hilal back to Hatzor where the aircraft was examined. It had received a pair of 23mm hits from the MiGs and three hits from small arms fire.

At 1130 hrs another pair of Ouragans were dispatched on an interdiction sortie to the Mitla Pass, and on their way home were bounced by two MiGs, of which only one pressed on with the attack and stayed with the Ouragans. This time the Ouragan pilots were more aggressive. With more fuel and facing only a single MiG, the leader, Jacob Agasi, took advantage of the fact that the Egyptian pilot was chasing his wingman alone. The wingman dragged the MiG into a slow combat of tight turns in which it lost its advan-

The Meteor was fitted in Israel with an improved cockpit canopy which provided better visibility for the pilot. (IAF Magazine)

tage of higher top speed and superior climb rates. Moreover, the MiG was all alone without a wingman to clear his tail, and the Egyptian pilot concentrated completely on the Israeli wingman, allowing Agasi to creep slowly into firing position. He missed the first burst, but in the second the MiG received several hits from his four 20 mm guns in the left wing root and the tail section. It lost a few pieces of metal, fuel leaked from the left wing, and the engine poured black smoke. The MiG immediately headed west, although Agasi was able to fire yet another short burst in which a few more hits were scored in spite of the fact that only a single gun fired. The faster MiG rapidly disappeared from Agasi's view while poring black smoke, and he was credited with a probable kill.

During noon the Harvard squadron also got back in action, and it launched the first formation to attack Egyptian vehicles on the El Arish-Abue Ageila road at 1250. The Mustangs were also very busy on CAS and interdiction missions, and during one of those the commander of 105 Squadron at Ramat-David, Rav Seren Moshe Tadmor, was killed when he climbed to height in contrary to the regular practice to fly as low as possible, and his Mustang was struck by flak at Abu Ageila. Seren Avraham Hacohen was supposed to succeed Tadmor, but since he was ill Seren Shlomo Hagani, a reservist, took office as the new CO of 105 Squadron. During the same noon hours a fierce battle was taking place at the Mitla Pass as the 202nd Brigade, already linked with the 890th Regiment, commenced an attack to completely occupy the pass, which was blocked by an Egyptian regiment, and during the seven hours of combat some 40 Israeli paratroopers were killed against the 260 bodies of Egyptian soldiers found in the area. The Egyptians received an efficient aerial support in the shape of four Meteors escorted by six MiG-15s, which attacked the Israeli forces at about 1600 hrs that afternoon. Although several Ouragans were in the area, the EAF attack was not disrupted and the paratroopers suffered hard. Seven were killed and another twenty injured, while a 120mm mortar, an ammunition truck and three vehicles were destroyed.

Very rare photograph of a Nord Noratlas (4X-ALH) in El Al's colors at Sharm el Sheikh on 6 November 1956. It was probably a French aircraft on loan to Israel, although it is painted as a civilian aircraft and carries the Sinai Campaign identification stripes.(GPO)

Four of the six Ouragans that were flying over the Mitla Pass area at the time of the EAf attack, but were unable to intercept the Egyptian raiders, met a pair of MiG-15s on their way back, but the two formations merely passed one another and no air combat emerged. The air activity in the afternoon slowed down somewhat, and the Ouragan squadron mounted ten sorties versus a total of 48 for the whole day, although for the first time during the Sinai Campaign, 101 Squadron Mystères landed at night when six of them returned to Hatzor after darkness fell.

Wednesday 31 October was to be the first full day of fighting in which the Chel Ha'avir would enjoy total air superiority thanks to the Anglo-French attack on the EAF. However, that attack was postponed from 0600 that morning to 1900 in the evening. As a consequence, the IDFAF had no option but to operate its piston fighters under the threat of Egyptian interceptors. In reality, it turned out that the flak was the deadliest enemy; three Mustangs and a Harvard were lost to that cause on the day. A Mosquito PR.16 was also lost on this day when it crashed at Tel-Nof while returning from a reconnaissance mission, and the crew escaped unhurt. In the air combat arena the IDFAF was finally getting good results with six confirmed kills—four Vampires and two MiG-15s—in five recorded air combats, in which a single Ouragan had to land at Chlotsa Sands due to fuel starvation. The total number of sorties flown by the EAF that day was estimated by Israeli Intelligence at around 90, which was more that twice the previous day, but still a low figure comparable to the total number of sorties flown by the resident units at Hatzor, 101 and 113 Squadrons, which performed almost one hundred sorties that day. While the fighter pilots were working hard to support the IDF ground forces and keep the EAf away, the transport crews were briefed at 1600 hrs for yet another night of action, which, in addition to the dropping of water, fuel, ammunition and supplies, the Dakotas had the added task of landing at the Mitla Pass airstrip with reinforcements and to evacuate casualties. Just like the previous night they flew in singles at preplanned intervals. The first Dakota was unable to locate the paratroopers, but the second and third did, and after they dropped the supplies they were requested to land and evacuate the wounded soldiers, although the initial plan envisaged that other Dakotas who would also fly reinforcements later the same night would perform this task. Permission was granted, the Dakotas landed, and the evacuation of more than 120 casualties suffered by the paratroopers the previous day immediately commenced, with many lives saved thanks to 103 Squadron crews. Armée del'Air Noratlas from Cyprus also contributed to the effort, and ten of these dropped supplies at the Mitla Pass that night, although they missed Doron, a fact that gave rise to the speculation that the Egyptians were parachuting airborne forces in the area!.

Slightly to the north, other IDFAF piston types were operating to attack Egyptian units in northern Sinai. The B-17s were to perform their operational debut during the Sinai Campaign by bombing the Rafah area, and the plan was that two pairs of Harvards would follow. The bombing, executed between 0230 to 0305, was not successful, as the flares dropped by the old bombers illuminated the IDF forces in the area. Moreover, one of the Harvards was unable to locate its base, so his pilot dropped the flares, pro-

During 1956 the Meteors of 117 Squadron performed mainly in the air interdiction and close air support roles. Note the identification markings applied to the rear fuselage.(IAF Magazine)

viding the Egyptians the possibility to shell the Thsal (Israeli ground forces) in the area despite the black out. The activity of the EAF that night was not successful, either. An unknown number of Il-28s discharged their bomb load on fields near Gezer, east of Hatzor, without causing any damage. It was the second and last effort to bomb Israeli targets at night during the Sinai Campaign, because at 1900 on 31 October RAF Canberras and Valiants commenced their attack on the EAF air bases with the intent of destroying all aircraft and installations within 48 hours. The Egyptians immediately commenced the evacuation of many aircraft to their friendly allies, Syrian and Saudi Arabia, and to distant bases, and thus many were saved, although under the Anglo-French pressure the EAF completely disappeared from the skies of Sinai, and for the rest of the campaign the Chel Ha'avir enjoyed a total air superiority. Furthermore, the Egyptian armored forces were ordered to abandon Sinai and confront the Anglo-French forces that landed at Port Said and Port Tauffic on the northern entrance to the Suez Canal, leaving less forces in Sinai to tackle the IDF ground forces.

With no air threat the IDFAF concentrated during the last five days of the Sinai Campaign on CAS and interdiction. The Harvards were withdrawn from operational use on the morning of Thursday, 1 November, due to their high vulnerability to flak and their general ineffectiveness. The unit, 140 Squadron, executed a total of 24 sorties in which 82 hours were flown and used 320 hollow-charge 80mm rockets and 1,000 12.70mm rounds. While the Harvards ended their part in the Sinai Campaign and were returning to Tel-Nof, the Armée de l'Air Mystère and F-84F Thunderstreak units at Ramat David and Lod* started flying interdiction missions against the retreating Egyptian forces. The Mystères flew a total of 92 sorties and attacked vehicles on the El Arish-Quantara and the Abu Ageila-Ismailiya roads in conjunction with the Anglo-French action on the northern sector of the Suez Canal.

On Thursday, at least three Mustangs crash landed due to the dangerous flak, but all pilots were recovered. That day the Mosquitos commenced operations in the daytime, and their primary target was Ras Nasrani and Sharm el Sheikh, to which the 9th Brigade commenced its advance that evening to achieve one of the original goals; to open the Strait of Tiran to Israeli commercial shipping. The 9th Brigade waited at Eilat until total air superiority was achieved as it had to advance through some 200 Kilometers of very difficult terrain. To aid the brigade in its movement, 100 Squadron dispatched several Piper Cubs to Eilat that constantly flew ahead, transmitting back their findings on the radio.

No air opposition was encountered during Thursday, and the only operational activity on the night of 1/2 November was the flying of fresh units and supplies to the Mitla Pass by 103 Squadron Dakotas. Friday, 2 November, witnessed the collapse of the Egyptian army network in northern Sinai and the breakthrough of the IDF armored forces towards the Suez Canal. The Mustangs that supported the 37th Brigade in its attack on Om Kettef reported that the Egyptian strongpoints were deserted. That report was indeed correct, but the 37th Brigade advanced too quickly and ran into an ambush from the IDF's 7th Brigade on the western side of the deserted Egyptian array. A pair of Meteors noticed the mistake and dived between the fighting forces, but were not understood. The fire continued until a Piper Cub pilot, Paltiel Sikortin, arrived at the scene and landed near the 7th Brigade forces. He explained to them their mistake and then flew between the forces and directed them until they met. Although Sirotkin was decorated, his action came too late and a complete 37th Brigade company was wiped out in that confused "own goal" combat.

While the 7th and 37th Brigades were clashing west of Om Kattef, the 27th Brigade was advancing further north and captured El Arish air base, which was immediately used by the Piper Cubs that were supporting the various forces in northern Sinai. The focus by then was turned to the 9th Brigade, which was facing severe difficulties in its effort to overcome the unfriendly terrain along the eastern shores of Sinai. To speed up the capture of Sharm el Sheikh, air attacks were flown against the Egyptian forces in the area, which were well defended against air raids, with a few German WWII 88mm and dozens of 30 mm AA guns, which damaged some Chel Ha'avir aircraft in the coming days and even destroyed two of them. In addition to the air attacks it was decided to drop two forces of

*Disguised as 201 and 200 SQNs respectively.

paratroopers; one at Sharm el Sheikh itself and the other at A Tur on the western coast of Sinai, some 70 km north of Sharm el Sheikh, where a coastal road existed. This was not the case with the eastern coast, and thus the slow advance by the 9th Brigade. In this way the paratroopers were to advance from A Tur south to Sharm el Sheikh to speed up its capture.

Almost every type of Chel Ha'avir combat aircraft took part in the effort to soften the Egyptian forces at Ras Nasrani and Sharm el Sheikh in spite of the difficulties. Both targets were relatively small and well defended against air raids, and the reasonable approaches to attack were north to south or vice versa, as the targets were located between the Gulf of Aqaba and the mountainous central Sinai. It was possible to bomb from high altitude, but such an action was less accurate and still quite dangerous as witnessed by the pair of B-17s that were sent to bomb the 130 mm coastal guns at Ras Nasrani and the flak positions at Sharm el Sheikh. Both were flown by veteran ex-Commanding Officers of 69 Squadron: Jacob Ben Chaim, who was the then CO of the reborn unit, leading the squadron during 1950-1952, while Nachom Efrat succeeded him and led the unit during 1952-195. Efrat bombed his target and escaped unscathed, but Ben Chaim's B-17 suffered a direct 88 mm hit (shades of WWII!) in its left wing. He immediately jettisoned all the bomb load, shut down one engine and headed north while barely controlling the vibrating bomber. In spite of all that the damaged B-17 regained its home base, Ramat-David, where a large hole was found in the left wing during the post flight inspection.

Another squadron commander who returned from Sharm el Sheikh with a hole in his left wing was Itschak Evene, whose Mustang received a direct hit, but he was able to regain Tel-Nof. Less fortunate was a third CO, Benjamin Peled of 101 Squadron. Two formations of Mystères were dispatched to the area: one to Ras Nasrania and the other to Sharm el Sheikh. Peled led the four Mystères to Ras Nasrania. Although only three arrived at their assigned target, Ran Peker had to abort north of Eilat due to a malfunctioning fuel system. During the second or third pass Peled's Mystère was hit by flak, when he was then seen to pull up his damaged fighter, which pored black smoke, to 5,000 ft and eject, receiving the dubious honor of being the first Israeli pilot to use an ejection seat. He landed some three kilometers from the Egyptian strongpoints and immediately ran west toward the mountains to evade capture. Several formations of Mosquitos and Mustangs were ordered to cover his estimated position and attack any Egyptian search party, and indeed a Mosquito formation successfully attacked a pair of armored vehicles which were probably looking for the Israeli pilot. While all the Mosquitos returned to their base, the Mustangs were less fortunate and one was lost to flak. Yonnatan Atkes was seen to make an emergency belly landing north of Ras Nasrani, while his wingman, Eliezer "Chita" Cohen, stayed in the area and the other two Mustangs returned home, reporting that Atkes left his Mustang and immediately laid down making no effort to escape the Egyptians which came and took him as the only IDFAF POW during the Sinai Campaign. Later it was revealed that Atkes was injured and lost a lot of blood, and for that reason he landed—otherwise his Mustang was flyable. Atkes was shipped to Egypt from Sharm el Sheikh before the area was captured by the IDF, and thus he spent several unpleasant weeks before being exchanged with the 5,500 Egyptians captured in Sinai by the Israelis. He was freed on 31 January 1957.

While the Mustang and Mosquitos were covering Peled's position a Piper Cub was sent from Eilat to recover the missing 101 Squadron CO. Arriving at the area at 1500, some two hours after he had ejected, the Piper could not locate his position. Another Piper Cub was at that time supporting the 9th Brigade, and it was also ordered to abandon its observation mission and join the first Piper in its search. Avraham Greenbuem, who flew that Piper, climbed slowly to 10,000 ft to avoid flak, and headed south. North of Ras Nasrani he started to descend, and at 4,000 ft discovered a big black crater that might have been caused by an aircraft upon crashing, so he commenced a search pattern in the surrounding area in which two Egyptian soldiers were seen. Greenbaum's observer was ready to shoot at them with his Uzi submachine gun, but the pair ignored the descending Piper, so the Israelis held their fire. It was almost 1700 hrs and the sun was low, but the pilot and observer suddenly saw a lonely man waving a white piece of cloth. The two were afraid of an Egyptian trap, so they headed towards the man with the intention to identify him first. As already mentioned, the Chel Ha'avir was a small organization at the time and everyone knew everybody. During the first two low passes Greenbaum was not

The Sinai Campaign was the swan song of the Mosquitos in 110 Squadron service; soon after they would be replaced by Vatours.(GPO)

sure that it was in fact Peled, but on the third, all doubts left him and he immediately landed in a dry wash a few dozens of meters from Peled's position. However, he was unable to move because of a broken ankle, and the fact that in spite of that injury he escaped towards the mountains a few hours earlier worsened his pains. Greenbaum and his observer left the Piper and carried the wounded Peled to their aircraft. The door of the Piper, which was not designed to carry three men, barely closed, but the faithful Cub did not disappoint the rescuers or the rescued, and the flight to Eilat was uneventful. Once there Peled was transferred to a Dakota and flown to Tel-Nof, where he was greeted by his commander, Ezer Weizman.

While the rescued 101 Squadron CO was flown north two formations of Dakotas were heading south for an airborne assault on Sharm el Sheikh and A Tur. Two companies were to be parachuted, one at each target, at dusk, and the seven Dakotas left Tel-Nof at 1530 hrs. However, by then it was well known that the flak at Sharm el Sheik was very dangerous, and it was decided to avoid risking the slow and low flying Dakotas and parachute the two companies at A Tur. When that decision was taken the Dakotas were already on their way and the attempts to inform the formation flying to Sharm el Sheikh about the change failed because the Dakotas flew at an altitude of some 300/500 ft. A Meteor was instructed to catch the Dakotas and indeed the message was delivered some twenty minutes before the estimated time on target. The Dakotas then headed to A Tur from Sharm el Sheikh, but were too late to parachute in the daytime, so this operation was made under the light of torches. One of the Dakotas landed at A Tur, following the plan established to check the landing strip there. After it an air shuttle was organized—after the paratroopers cleared the area and secured the landing strip—in which a regiment of the 12th Brigade was flown in by the Dakotas and Noratlas, which made 23 round trips, and a single El Al Constellation, which was recruited for the operation and made two round trips that night.

Also on the afternoon hours of 2 November the 890th Regiment commenced to advance from the Mitla Pass towards Ras Sudar, some 160 km north of A Tur on the Gulf of Suez shores, aided by an observation Piper Cub reporting that Ras Sudar itself was deserted.

With total air superiority in hand it was only natural that on Saturday 3 November, a Dakota flew the IDF CoS Moshe Dayan, the General Director of the Ministry of Shimeon Peres ,and the poet Nathan Alterman all over Sinai in the daytime and without any escort. Moreover, the pilot was none other than Yael Rom. They landed in the morning at A Tur, and from there continued back to Israel with stops at the Mitla Pass, Bir Hama air base where the 7th Brigade advance HQ was located, and El Arish air base, from where they flew back to Ramle, arriving at 1900 hrs. The fighting that continued that day was primarily concerned with the advance of the IDF forces to Sharm el Sheikh from both available directions, the Gulf of Suez (12th Brigade) and the Gulf of Aqaba (9th Brigade). The Piper Cubs supported the advance, and during the morning Itschak Hirsch aided the 9th Brigade in avoiding an Egyptian ambush.

Towards the afternoon the forces were nearing the Egyptian garrison at Ras Nasrani and Sharm el Sheikh and aircraft were sent to soften the defenders. At 1600 hrs the first operational mission for that day by 116 Squadron at Tel-Nof departed the base to indicate the low level of activity and the fact that the peak was over. The Mustangs could not make contact with the ground forces, so they looked for targets of opportunity and discovered a warship sailing east of Ras Nasrani. Intelligence reports warned that the Egyptian frigate Damiat was about to reinforce Sharm el Sheik, and thus the Mustangs immediately came in from 12,000 ft. During a single pass, all except two released their ordnance—first rockets and then napalm bombs—while firing their 12.70mm guns. The pair that did not release its napalm bombs came in for a second pass. The flak was heavy and accurate, but all Mustangs regained their base while black smoke was seen to rise from the warship. However, none of the napalm bombs hit the evading ship, and although some of the rockets did, they caused no damage as they were not AP rockets. Nevertheless, the warship's troubles were not over, as it was also attacked by a quartet of Mystères.

Eight 101 Squadron fighters in two formations were sent to Sharm el Sheikk and Ras Nasrani that afternoon, but they were unable to locate new targets so, like the Mustangs, they turned their attention to the gulf of Aqaba. The first formation sank a small wooden boat off Ras Nasrani and then noticed the larger warship deeper into the sea. However, this formation reached bingo fuel, so they directed the second formation to that warship. In spite of heavy AA fire the four Mystères came in and each pilot fired his 38 rockets which, like those of the Mustangs, were not AP, as all aircraft in the area were to attack well-dug ground forces rather than warships, for which purpose high explosive or hollow charge rockets are better suited, and thus did not inflict severe damage. The one thing that all pilots were unaware of was that they were not attacking the Egyptian frigate Damiat, but the British cruiser HMS Crane, which had sunk the Egyptian warship on 1 November.

The next day, Sunday, 3 November, the 9th Brigade made its first attempt to assault Sharm el Sheikh, but failed in spite of the CAS provided by the Mustangs, whose pilots reported that the flak was no longer heavy nor too dangerous.

5 November was the last day of fighting as far as the IDF was concerned, and the first real day of fighting for the British and

After the conflict Mystères continued to be delivered from France, as seen by this machine just arrived from France in January 1957. (GPO)

The Ouragans proved to be realiable ground attack platforms, with a heavy punch from their rocket/bomb load and their four 20mm cannon.

French, whose forces landed that day, from the sea and from the air, at Port Said and Port Tasufik on the north of the Suez Canal. The Chel Ha'avir concentrated on that day in supporting the forces that finally overtook Sharm el Sheikh, thus ending hostilities at Sinai, and the limelight was turned to the Suez Canal Zone where the British and French were facing severe difficulties in their advance south. Except for CAS missions at Sharm el Sheikh, and the regular Piper Cubs which directed the forces from above, several Mystères were sent to the area on CAP after intelligence claimed that Saudi aircraft might attack the IDF forces, but this threat did not materialize. Once Sharm el Sheikh was captured the faithful Dakotas came in to bring fresh reinforcements, supplies and evacuate the wounded. From that day onwards until the Israeli withdrawal from Sinai, the nicknamed "Sinai Round Trip" was launched in which the Dakotas executed scheduled flights, operating almost like an airline, from Tel-Nof to El Arish, Bir Hama, A Tur, Sharm el Sheikh, Eilat and back to Tel-Nof

That day the Armée de l'Air F-84F squadron, which had arrived at Lod airport on 30 October, made its single most important contribution to the Sinai Campaign when, at the IDFAF's request, whose aircraft lacked the required range, attacked the EAF Il-28s that were dispersed at Luxor air base in upper Egypt on the Nile river, some 750 km from Lod, destroying 18 bombers on the ground while claiming two others as damaged, which wiped out the full inventory of the EAF bomber squadron based there.

The second Arab-Israeli war was over with a complete Egyptian defeat and the immediate threat on the actual existence of Israel removed. The British and French were less successful, however, and they had to halt their advance along the Suez Canal before they reached the southern end due to the pressures of the U.S. and USSR.

During the Sinai Campaign the Chel Ha'avir suffered from severe restrictions. It was unable to execute its intention to achieve air superiority by attacking the EAF at its bases due to political considerations. However, because of that fact it had to face the EAF for just two full days until the British and French commenced their attacks that rendered the EAF inoperable. Following this development, the IDFAF enjoyed the desired total air superiority over Sinai. During those days Israeli aircraft met EAF aircraft in the air 14 times. The Mystères were involved in eleven air combats, while the Ouragans fought two air combats and the Meteors only a single combat. Although the statistics claim that 262 CAP sorties were flown by the IDFAF, at least three of the combats involved aircraft that returned from CAS missions. On the Egyptian side, all combats except one involved the MiG-15s, while the exception was when an entire formation of four Vampires was wiped out by a pair of Mystères over the Mitla Pass. Initially, the Israeli Air Force claimed at least four MiGs and four Vampires, but that number was later reduced to only seven confirmed kills, two each by Yak Nevo, Yalo Shavit and Shai Agozi, and a single—in fact, the first was credited to Yosseph Tsuk. On the other hand, the EAF MiG-15s could only claim Benjamin Cahana's Piper Cub.

The lion's share of the total number of sorties flown by the Chel Ha'avir's combat aircraft during the conflict, 489 actually, were interdiction and CAS in which many Egyptian vehicles in Sinai were claimed as destroyed. During the first two days of fighting the air force was the only IDF arm to attack the main Egyptian armored forces, the 1st Armored Brigade and the 3rd Armored Regiment, with many vehicles claimed destroyed. To check these claims and verify which weapon caused the most damage a delegation of IDFAF pilots was sent on 4 November to Sinai to seek destroyed, damaged or abandoned Egyptian vehicles and investigate whether they were destroyed due to air power action. The conclusions were that of the limited number of weapons used the most effective against tank's armor were AP rockets. The 20 and 30mm rounds were unable to penetrate the armor of the tanks, while the napalm bombs were very effective, both physically and psychologically, against soft-skinned targets. Of the roughly 2,000 vehicles that the Egyptians operated in Sinai, about half were captured intact by the IDF ground forces, and of the rest the IDFAF was credited with 22 tanks, 17 armored vehicles, three half-tracks, eight Bren Carriers, four 40mm gun carriers and 254 other soft-skinned vehicles. Quite a respectable achievement, but the price paid was heavy, too: three pilots killed and twelve aircraft totally lost, including nine Mustangs, two Harvards and a single Mystère, with many other damaged during these ground attack missions.

The Sinai Campaign was also the greatest hour of the liaison pilots, who would be replaced in their traditional tasks in the future by helicopters. A total of 831 sorties were flown by the light-planes, an amazing average of about 103 per day, which represented 45 percent of the total number of sorties of all kinds, which was 1,846. The variety of missions fulfilled by the light-planes, especially the Piper Cubs (but also Boeing Stearmans) included: observation, liaison, commanders' transport, pilots retrieval and casualty evacuation. Three Piper Cubs were lost during the Sinai Campaign, of which two were attributed to EAF MiG-15s, while the third was lost on 6 November when Benjamin Gordon flew Aluf Asaf Shimchoni, the CO of the IDF Southern Command, back from Sharm el Sheikh to central Israel. The Piper probably drifted east during the long night flight and struck a mountain near Nablus, in Jordanian held territory, with both occupants losing their lives. IDFAF transports flew 192 sorties in which two airborne assaults were performed and 1,200 casualties evacuated from the frontline to hospitals in the rear. The remaining 42 sorties involved reconnaissance, in which a single Mosquito was lost in a landing accident, and the telephone lines' cutting by Mustangs which accounted for a total of 16 sorties.

One of the lessons learned by the air force from this war was the vulnerability of piston types to heavy flak, although in that context many sources claim that many Mustangs were lost in the Abu Ageila area since they were flying the same routes time after time, although Arie Dagan, who flew with 105 Squadron, claims that the routes were well planned and attributes those losses to the flak concentration at Abu Ageila, a key area over which most of the Mustangs and the two Harvards were lost.

On a higher level it was found that the command network was still lacking and the communications and intelligence gathering were still inadequate. It must be stressed that the IDFAF radio sets had only four channels, and thus all the airborne formations worked on the same frequency, thus overloading the network. Moreover, the codenames at the time were strict, and thus everybody airborne knew who was doing what and where at the time. Such a situation was far from today's trend. It was also a simple method that gave the enemy an easy life since all he had to do was to listen to a single channel. Thus, it was only natural that Segen Mishne Arie Dagan thought to attack his leader's Mustang, after he crash landed in Sinai, with rockets to deny the Egyptians from laying a hand on the radio frequency. The only reason that he did not attack was that he was not sure that Seren Eldad Paz had left his cockpit after the emergency landing near Bir Gafgafa on 31 October.

Nevertheless, and in spite of massive Armée de l'Air support, the Chel Ha'avir fulfilled its mission as defined by the IDF General Staff. Except for two unobserved raids, the threat of mass air attacks on targets within Israel failed to materialize, and the air force kept the Israeli skies open. In the CAS mission the scheme of attaching air support officers to each independent ground force proved correct as these officers, all pilots, knew exactly the capabilities and restrictions of aircrews and aircraft. The only mission that Aluf Dan Tolkowski so urged to perform, the destruction of the EAF on the ground at its home bases, had to wait for slightly over ten years.

OPPOSITE: The Chel Ha'avir during the Sinai Campaign.

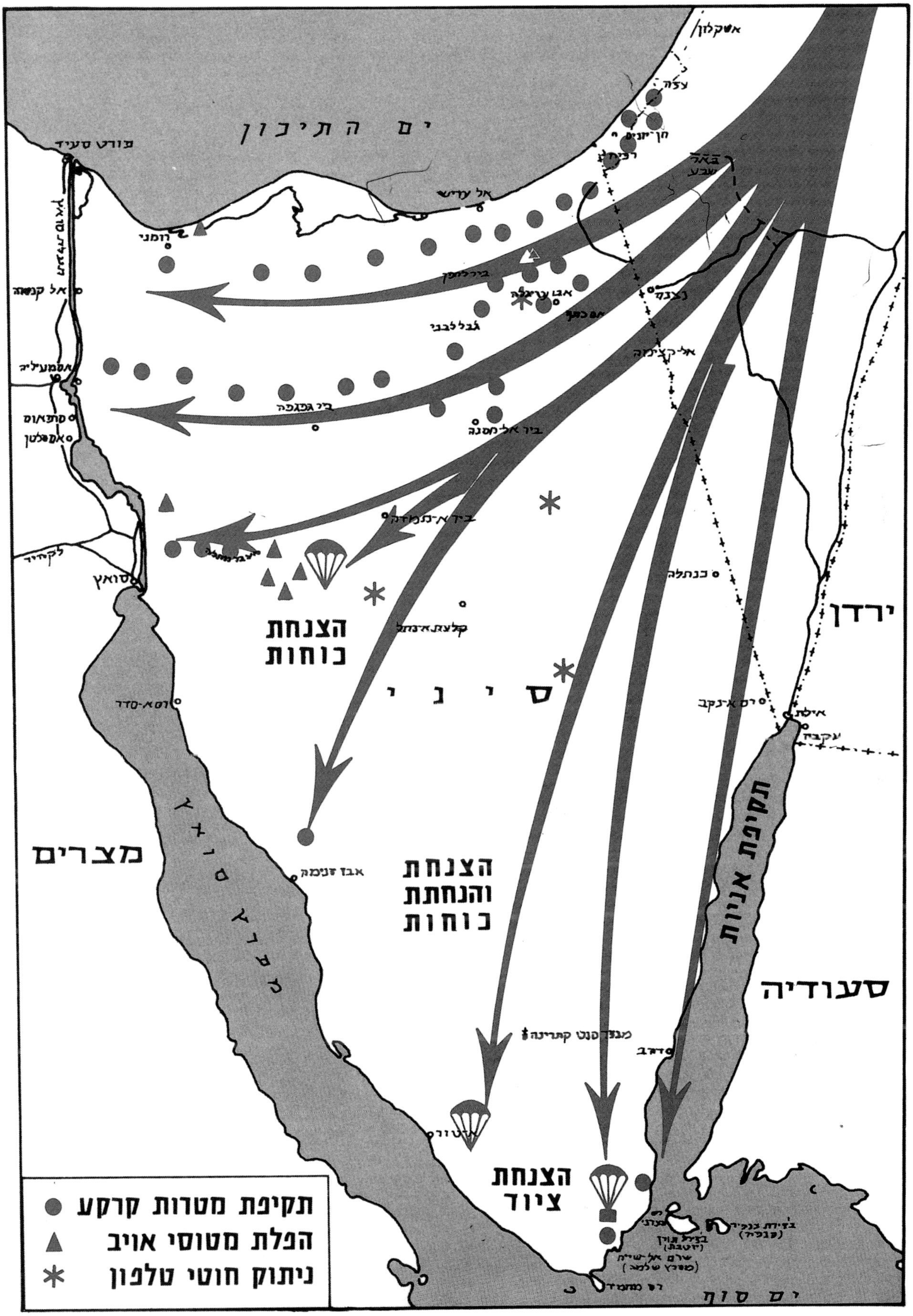
ים התיכון
פורט סעיד
אל עריש
אשקלון
עזה
באר שבע
ניצנה
ירדן
מצרים
סיני
סעודיה
הצנחת
כוחות
הצנחת
והנחתת
כוחות
הצנחת
ציוד
תקיפת אניות
מפרץ סואץ
סואץ
אילת
עקבה
ים סוף
תקיפת מטרות קרקע
הפלת מטוסי אויב
ניתוק חוטי טלפון

Chapter 10: The First Supersonic Jets

On 18 December 1956, less than two months after the end of the Sinai Campaign, Aluf Dan Tolkowski presented to the Prime Minister and Minister of Defense, David Ben-Gurion, and to the IDF General Staff, headed at the time by Rav-Allof Moshe Dayan, an ambitious re-equipment scheme that was to turn the IDFAF into an all-jet fighting force.

The performance of the Chel Ha'avir during the Sinai Campaign was beyond any previous anticipation, and its achievement moved Dayan to announce that from now on that the highest priority within the IDF would be given to the air force, even at the expense of the elite armored forces. Thus, Tolkowski had every reason to be sure of himself and optimistic when presenting his 93.7 million dollar plan to the General Staff. But then the gloom of reality once more interfered, and Ben-Gurion shortly concluded the meeting:

"I told the present that it is out of the question. I do not think that there will be a war this year (although skirmishes might occur with Egypt due to Sharm El Sheikh). I do not see the necessity of more Mystères. We should buy newer types. There is no necessity in eighteen Super Mystères and we can do with less than twelve Vatours. This year we will face positive immigration, and of the kind we have not seen yet: from Eastern Europe, thus Chel Ha'avir Staff should re-examine its requirements."

So, with major political difficulties and severe economical restraints, although the Air Force emerged from the Sinai Campaign as an efficient force and for the first time was truly appreciated by the IDF ground and naval forces of the IDF, it had to draw back its demands. Naturally, there was a continuation, and most of the plans bore fruit after the conflict, but almost as important as the possibility of acquiring new equipment was the level of experience acquired.

The most valuable gain of the Sinai Campaign was the operational experience achieved by both the then current leaders and the younger pilots—the future leaders. Also, it showed first-hand to the IDF General Staff the capabilities of modern, motivated, highly trained, and well organized air power. From that operational experience many lessons were drawn, and some of those lessons were to be realized during the late 1950s and early 1960s. The combat experience gained by Chel Ha'avir pilots during the Sinai Campaig was to bear fruit in the coming years with many of its young jet fighter pilots taking leading positions within Chel Ha'avir. The next five IDFAF commanders all flew operationally during the Sinai Campaign, and although this brief operational experience was later to be superseded by more extensive and intensive operational experience, it nevertheless was the first counting impression to many young combat pilots. It must also be remembered that many of the jet fighter pilots were very young and inexperienced by any standard. For example, the pilot credited with the first Mystère kill, Segen Yosspeh Tsuk, had less than twenty hours on the type when the Sinai Campaign broke-out, while Segen Ran Peker joined 101 Squadron only in September 1956, a month before the break of hostilities. More comprehensive statistics can be found in Ben-Gurion's diary: of the twenty Mystère pilots that took part in the Sinai Campaign, Ben-Gurion received from Tolkowski the total number of jet flying hours—Meteor, Ouragan and Mystère—at the time of the War for nineteen of them, and these figures are the following:

NUMBER OF FLIGHT HRS	NUMBER OF PILOTS
1,200	1
850	1
750	2
700	1
430-520	5
220-230	6
140	3

Thus, there can be no doubt that the impression that such an operational experience, gained so early in the flying career of the younger combat pilots, was indeed excessive.

It was also a matter of confidence, and the fact that seven Egyptian jet fighters were claimed in air-to-air combats against the loss of a single Piper Cub surely added to the confidence of the pilots in their own flying abilities, the tactics fielded, and the equipment used. One of the outcomes of that experience was a book about tactical air warfare written by Seren Ya'acov "Yak" Nevo, who scored two kills during the Sinai Campaign and was one of the 101 Squadron flight commanders immediately after the war.

The recent conflict also left its mark on personnel, and several changes were made shortly afterwards. The most significant of them was the appointment, on 1 December 1956, of Aluf-Mishne Ezer

This 1960 formation fly-past shows the then four jet combat types in Israeli service: Ouragan, Mystère, Super Mystère and Vatour. (IAF Magazine)

Weizman as head of Air Department within the IDFAF Staff, which was in fact the unofficial post of vice-commander. Although that nomination was agreed upon even before the Sinai Campaign, its importance cannot be overestimated since it marked the decline of Aluf-Mishe Shlomo Lahat's aspiration to succeed Dan Tolkowski as the air force C.O.. Lahat was the previous head of the Air Department and the leading contender to the post of IDFAF commander. Yet another significant appointment was that of Sgan Aluf Mordechai Hod, promoted from Rav-Seren, as the Hatzor air base commander in succession to Ezer Weizman, pioneering the trend that was to become clear in the following years and in which Hod stepped almost exactly on the footprints of Weizman until replacing him as the Chel Ha'avir's commander in 1966.

On the operational level the Sinai Campaign illustrated well the facts that the radar coverage and ground control network were still inadequate; that the best way to use the available force was by central control; that Chel Ha'avir still lacked a true all-weather capability; that the piston-engined combat types were outdated; and that some of the tactics used were also outdated. Of those five invaluable lessons, three concerned equipment, while the other two concerned tactical thinking; thus, it was easier to deal with the last two first.

Since the War of Independence the pair became the standard combat formation used by the IDFAF, the attempts to operate during the War of Independence in larger formations of four fighters proved fruitless because of the low serviceability rates and the small number of available fighters, although on the last day of the conflict with Egypt, 7 January 1949, a formation of four Spitfires did manage to get into the air and it was able to successfully engage a numerically superior formation of RAF Tempests, which had superior performance when compared with the Merlin-engined Spitfire, because of three main elements: initiative, combat experience and confidence. That success was not pressed on because seven years later, during the Sinai Campaign, it was not the number of serviceable aircraft, but the number of available jet fighter pilots that limited formation numbers; only 47 Israeli jet fighter pilots took part in this conflict, and in the early stages of the war the usual formation was still the pair. However, combat experience soon revealed that a formation of four fighters, comprised of a pair of elements of two each, was better suited to meet the Egyptian MiGs that flew in larger formations. The formation of four fighters was adopted by the IDFAF and indeed, during the Six Days War of June 1967, most of the formations were of four fighters, although for Quick Reaction Readiness (QRA) and Combat Air Patrol (CAP) the pair was retained.

The next operational lesson was the direct and central control of all Chel Ha'avir operations from a central operations room manned by Chel Ha'avir Staff officers. Such a structure necessitated several preliminary conditions: a reliable communications network from the central operations room down to squadron level; adequate ground control so the operational officers within the IDFAF staff would clearly see the current battle scene and thus react properly; and the ability of each air base to operate independently and support its fighting elements, the squadrons. The third lesson, which was closely related to the second, was the need of more extensive radar coverage. Shortly before the outbreak of the Sinai Campaign, the IDFAF received its first modern 3-D air radar from France, but it took several years before a modern radar network was established with American equipment. The honor of creating that radar network fell into the hands of Ezer Weizman, and was named YABA, the acronym in Hebrew of Yechidat Bakara Azorit (Sectorial Control Station), and two such stations were erected during Weizman's tenure as Chel Ha'avir commander: one in the Gallilee to cover Northern Israel and the second in the Negev to cover southern Israel, thus giving a complete radar coverage of the whole country.

The fourth of the main lessons was the relative lack of a true all-weather capability. It is true that prior to the Sinai Campaign a Meteor NF.13 downed an Egyptian Il-14 in a night action,. but the night fighter unit, 119 Squadron, was stood down for the duration of the Sinai Campaign, and although the two EAF Il-28 nocturnal attacks caused no damage, they were not met by any resistance. Moreover, the jet fighters were not operated after dark, although one evening several Mystères and Ouragans returned to Hatzor air base after darkness fell, but such an incident was the exception and not the rule. The night attacks that were carried out by the IDFAF during the Sinai Campaign, by the Mosquitos, B-17s and Harvards, used illuminating flares to help the pilots identify and attack their assigned targets. Although all weather capability was indeed not satisfactory, the best solution available was found with the activation of 119 Squadron, Ta'yeset Ha'atalef (The Bat Squadron), as a dedicated night fighter unit based at Tel-Nof air base under the command of Rav-Seren Yoash Tsidon, firstly equipped with Meteor NF.13s, with Vatour IINs added to into the inventory later.

Last of the main lessons was the unsuitability of the piston-engined combat types in the modern battlefield, and their phasing out of service commenced immediately after the end of the Sinai Campaign. The Mosquito fighter-bomber unit, 110 Squadron, was transferred from Ramat-David to Hatzor and disbanded there in early 1957 when the tension on the borders was relaxed after the U.N. sponsored Israeli withdrawal from the Sinai, completed in

March 1957, and the positioning there of an international U.N. peace-keeping force. Simultaneously, 69 Squadron, the B-17 unit, and 105 Squadron, the Mustang unit based at Ramat-David air base, were also disbanded, and thus by early 1957 Chel Ha'vir ORBAT was as follows:

HATZOR	TEL-NOF	RAMAT-DAVID
101 Sqn. (Mystère)	115 Sqn. (Mosquito PR16)	109 Sqn. (Mystère)
113 Sqn. (Ouragan)	116 Sqn. (Mustang)	117 Sqn. (Meteor)
119 Sqn. (Meteor NF.13)		
103 Sqn. (transports+helos)		
12 Flying School		

To this order 100 Squadron at Ramle air base with its liaison and observation duties must be added, while 109 Squadron was reformed at Ramat-David air base as a second Mystère unit using the large number of available Mystères that were flown during the Sinai Campaign by Armée de l'Air pilots as "201 Squadron" due to pure lack of Israeli jet fighter pilots. 109 Squadron was reformed as a Mystère unit under the command of Rav-Seren Menachem Bar on 16 December 1956.

The phasing out of service of the old piston combat aircraft resulted in several changes and new demands. One of those new demands was for a new medium jet fighter/bomber as a direct replacement to the veteran Mosquitos. As early as September 1953 Aluf Tolkowski recommended the procurement of English Electric Canberras, but political considerations resulted in the acquisition of a roughly equivalent French design, the SNCASO SO.4050 Vatour twin-engined jet fighter/bomber.

The Vatour was faster than the Il-28 and could carry up to 3,500 kg of bombs to targets that were previously beyond the reach of Chel Ha'avir's combat aircraft—except for the venerable B-17s. The economical difficulties resulted in a lengthy controversy as to the exact numbers of Vatours to be procured. The initial contract was signed in August 1956 and was comprised of twelve Vatour IIAs and six Vatour IIBs (that were later canceled and exchanged with Vatour IIAs), each costing 742,850 dollars, excluding spare parts and ammunition. Four months later, during the IDF General Staff meeting of 18 December 1956, Aluf Tolkowski demanded another dozen Vatours, but Ben-Gurion rejected this proposal on budget grounds. On 1 August 1957 the first Vatour arrived at Hatzor air base, while the second arrived a fortnight later, but just like the first and the others that were delivered in the coming months, was stored at this air base since the French were reluctant to reveal to the world that they supplied Israel with such an offensive weapon system.

Later on the Vatour IIAs were organized in a single unit, 110 Squadron, at Ramat-David air base under the command of Rav-Seren Yechezkel Somech, who was the last commander of 110 Squadron in its preceding form as a Mosquito unit. Two other subtypes of the Vatour family were also to enter IDFAF service in the coming years: the Vatour IIBR reconnaissance version were to be operated by 110 Squadron, thus allowing the disbandement of the Mosquito PR.16-equipped 115 Squadron at Tel-Nof, while the Vatour IIN night fighters were to augment 119 Squadron's Meteor NF.13s at the same air base of Tel-Nof.

The charismatic Ezer Weizman became the Chel Ha'avir's C.O. after the Sinai Campaign and up to 1966, then reponsible for the acquisition of the Mirage III Mach 2 fighters.(GPO)

The pilot that ferried the first Vatour to Israel was Danni Shapira, a natural pilot by birth, who became the first IDFAF Chief Test Pilot (CTP) after graduating from the French test pilot course at Bretigny and Istres. Another graduate of the same course was Hugo Marom, who was sent by Israel Aircraft Industries.

Other combat pilots also enjoyed the warm relationships with the Armée de l'Air when single direction, from Israel to France, exchange tours were introduced and many Chel Ha'avir pilots served up to a year with regular French units, gaining valuable experience in the process. On 27 June 1957 it was reported at the weekly IDF General Staff meeting that 24 Israeli pilots were already training in France. Among the pilots that were sent to France and benefited from that arrangement were Asaf Ben-Nun, Arie Dagan, David Kishon, Gideon Maggen, Yak Nevo, Ran Peker, Elan Ron, Levy Tsur and others.

Arie Dagan and Gideon Maggen were among the group of six 109 Squadron pilots that were sent to France in May 1957. They both claim that the pilots were sent to fly in France because budget cuts dictated a reduction in available flying hours in Israel. They flew in a regular Armée de l'Air (2e EC) Mystère unit at Dijon, but accumulated only sixty flying hours during the six months they spent in France. The most valuable lesson they learnt in Europe was flying in bad weather—IFR flying in Israel at the time was not encouraged. Maggen also recalls that they flew training missions against the air forces of several NATO members on a regular basis and always lost to the RCAF pilots that flew the Sabre.

As already noted earlier in this narrative, David Ben-Gurion, the Prime Minister and Minister of, rejected the procurement of more Mystères since newer types were offered to Israel and he felt that it only made sense to buy the most advanced types that could be procured. In the years that immediately followed the Sinai Campaign the most advanced fighter that Israel was able to acquire was the Dassault Super Mystère B.2.

The Super Mystère B.2, or SMBD as it was often called, was the first level supersonic fighter to enter Israeli service. It was also the first European supersonic fighter to enter service. Although it was smaller, lighter and generally inferior to the North American F-100 Super Sabre, there can be no doubt that it was inspired by the successful American design.

As early as September 1955 the French offered Israel eighteen Super Mystères, but it was clear at the time that it was still merely a promising prototype that first flew only six months earlier. Finally, in November 1956, Israel ordered 24 Super Mystèteres even before an Israeli pilot evaluated the type!. IDFAF CTP, Danni Shapira, first flew the new supersonic fighter on 26 February 1957. In anticipation of the arrival of the new fighter 105 Squadron, Tayeset Ha'akrav (The Scorpion Squadron), was reactivated at Hatzor air base under the command of Rav-Seren Yak Nevo, with Seren Zoric Lev and Seren David Ivry as his flight commanders. A small group of pilots was sent for a short conversion course at France in September 1958, and three months later, on 3 December 1958, the first SMBDs arrived to Israel. As 105 Squadron became operational towards early 1959 the Mystères lost their natural metal finish in preference to the then standard camouflage of brown and blue upwards and light sky downwards. Up until then 101 Squadron Mystères could have been identified by a narrow red line running along both sides of the fuselage, while 109 Squadron Mystères were decorated with a red lighting running along both sides of the fuselage. Once camouflaged the Mystères lost such identities and only the unit badge was painted on the vertical fin. On the other hand, the new Super Mystères were left, according with the then current tradition, in their bare metal finish and received a stylish red lighting on both sides of the fuselage, while also enjoying their unit badge painted on both sides of the vertical fin.

During the same time scale another important addition to the IDFAF Order of Battle was the formation of the first helicopter unit, 124 Squadron, which was nicknamed Tayeset Ha'rotor Ve'hacherev (The Rotor and Sword Squadron). That unit saw its origins in 103 Squadron Helicopter Flight, which was created in November 1956 after the pair of Sikorsky S-55s arrived from the U.S. in crates during the previous April. The S-55s were ordered in December 1955 for 600,000 dollars, and two pilots, Rav-Seren Yehuda Arbel, 103 Squadron Commander, and Seren Ori Yarom, the intended commander of the yet-to-be-formed Helicopter Flight, were sent for conversion in the U.S. in July 1956. As they returned in November 1956 the new Helicopter Flight was formed within 103 Squadron at Tel-Nof air base, and the lonely Hiller 360, that was rarely flown by 100 Squadron, was transferred to the new Helicopter Flight.

During the weekly IDF General Staff meeting of 18 December 1956, Aluf Tolkowski demanded one million dollars for the procurement of three Sikorsky S-58s, but that demand was revised the following summer after a delegation of IDF officers toured France and Algeria to learn first hand about the use of heliborne forces. The delegation consisted of three ground forces officers at the rank of Sgan Aluf—they were in fact the head of the Armament Branch and the head of the Tactics Branch within the IDF General Staff and the Southern Command Chief of Staff, and the commander of the Helicopter Flight, Seren Yarom. When the delegation returned to Israel in February 1957 it recommended the procurement of Sikorsky S-58s and Alouette IIs, and as a result IDFAF requirements were altered to include the French type also. In the 21 June weekly General Staff meeting Tolkowski demanded one and half million dollars for the acquisition of three S-58s and three Alouette IIs. It seems that the procurement of the S-58s was approved while that of the Alouettes was rejected, and in February 1958 the three S-58s arrived by ship to Israel. However, by this time the Helicopter Flight had already turned into 124 Squadron and operated a single Alouette II, also.

By the early summer months of 1957 a wealthy French widow, Madame Ductz Lammaret, donated the IDFAF a single Alouette II under the terms that it would only be used for mercy flights, and once that term was accepted a Chel Ha'avir Nord Noratlas collected the new acquisition and flew it to Tel-Nof air base, where it arrived in July 1957. The addition of the Alouette and the anticipation for the arrival of the S-58s raised a justified claim to separate the Helicopter Flight from 103 Squadron, and on 1 January 1958 the new unit, 124 Squadron, was officially formed with RAv-Seren Ori Yarom as its first commander. During the coming years, 1959 and 1960, the squadron received another four S-58s, and by the end of the fifties it had seven S-58s in addition to the two S-55s, a single Alouette, and the little-used Hiller.

Up to the end of the decade other minor additions to the ORBAT included three more Meteor NF.13s to augment 119 Squadron, and six of the ten C-47s that were loaned to the IDFAF by the Armée de l'Air during the Sinai Campaign were retained by 103 Squadron, which also benefited from the appreciation that the IDF General Staff felt towards the Noratlas. Small numbers of the twin-boom French transport were constantly added to 103 Squadron inventory until a total of 24 were delivered by the early sixties.

Thus, as the Chel Ha'avir entered its second decade of existence, its ORBAT almost entirely consisted of jet combat aircraft:

HATZOR	TEL-NOF	RAMAT-DAVID
101 Sqn. (Mystère)	103 Sqn. (transports)	109 Sqn. (Mystère)
105 Sqn. (SMBD)	116 Sqn. (Mustang)	110 Sqn. (Vatour)
113 Sqn. (Ouragan)	119 Sqn. (Meteor NF13 & Vatour IIN)	117 Sqn. (Meteor)
124 Sqn. (helicopters)		
12 Flying School		

Of this ORBAT, 117 Squadron became the advanced stage of the Flying School. As is explained later, 113 Squadron became the Operational Training Unit (OTU), and 116 Squadron was an emergency unit which consisted primarily of reserve pilots that were not converted to jets. Another unit, 100 Squadron, was transferred during 1959 from Ramle air base to Sde-Dov air base on the northern suburbs of Tel-Aviv, the very place where 100 Squadron's parent unit, the Tel-Aviv Squadron, was first organized during December 1947.

The move of 100 Squadron from Ramle air base, which was handed over to the ground forces, was dictated by the move of the IDFAF Staff from its Ramle camp to Tel-Aviv in an adjoining group of buildings to that of the IDFAF General Staff. The move was affected in April 1959 and was marked by a fly-past of six Mystères over the new staff camp.

The late fifties also saw three independent moves that were to culminate in the public image, for good, of the Chel Ha'avir image. Those three steps were the replacement of Aluf Dan Tolkowski as its C.O. by Aluf-Mishne Ezer Weizman in June 1958, the inauguration of the by-now traditional IDFAF major public fly-past the same month, and the reorganization of the Flying School. During the summer of 1958 the IDFAF celebrated its first decade of existence with an impressive fly-past and a demonstration of firepower in the Negev. For the first time the new French fighter bombers, the Vatours, were revealed to the public, and with all the other combat aircraft, from the Mustangs, via the Meteors and Ouragans to the Mystères, also taking part in the air display. It was also the first time that a formal aerobatics team, consisting of six Harvards flown by the Flying School instructors and led by Rav-Seren Israel Ben-Sahar, appeared before the public. It is true that previously, espe-

Formation of a Meteor NF.13, re-engined Super Mystère, and Ouragan.(GPO)

cially during the annual military parade on Israel's Independence Day, the air force organized fly-pasts like that of May 1957 in which 72 jet fighters overflew Tel-Aviv. Nevertheless, it was the fly-past in the Negev that marked the inauguration of the by-now traditional and highly publicized annual Chel Ha'avir Day.

The air display in the Negev was the swan song of Aluf Tolkowski's period of five years as the IDFAF commander, and the climax of that period was undoubtedly the Sinai Campaign. On 23 July 1958 Aluf Ezer Weizman took office as the new commander in a formal ceremony.

Aluf Ezer Weizman was born in Tel-Aviv in 1924. His uncle was Chaim Weizman, who was destined to become the first President of Israel, a post held today by none other than Ezer Weizman!. Weizman learned the trade of fighter pilot with the RAF toward the later stages of WWII. He graduated as a Flight Sergeant early in 1945 and proceeded to an OTU at Fayid with Thunderbolts. After flying and fighting in the War of Independence he was promoted to the C.O. of Israel's premier fighter unit, 101 Squadron, on 1 April 1949. He was—and still is—a fighter pilot by heart; he even claimed an RAF Tempest while flying a Spitfire with 101 Squadron in January 1949. His inspired leadership in his new post was the missing link that turned the air force into the spearhead of the IDF.

The third move to line up Chel Ha'avir with the then current state-of-the-art was the reorganization of the Flying School and its syllabus. As early as September 1956 it was decided to skip the OTU stage on Mustangs and convert the fighter pilots' stream of graduates directly to the Meteor. Up until then that syllabus consisted of a primary stage on Boeing-Stearman PT-17 Kaydets and a basic stage on North American T-6 Harvards, after which the cadet received his wings, and the rank of Segen-Mishne, before proceeding to Ramat-David air base to attend the OTU, 105 Squadron equipped with Mustangs. As a result of the decision to drop the Mustang stage, it was only natural to disband 105 Squadron, but the Sinai Campaign delayed both the change of syllabus and the disbandement of 105 Squadron. However, as the war ended both actions were taken and the new syllabus consisted of the primary stage on Stearmans, the basic stage on Harvards and, instead of receiving wings, the fighter stream cadets were moved to Ramat-David air base to complete the newly introduced advanced stage on Meteors, for which purpose another two dual-seat Meteors were acquired in Belgium. These were Meteor T.7/8s because they had the T.7's fuselage mated to the F.8 tail section. Only upon the successful completion of the advanced stage with 117 Squadron did the pilot receive wings and move on to Hatzor air base to attend the new OTU, 113 Squadron on Ouragans.

However, more radical changes were in the pipeline. Several of these concerned the Flying School itself. The custom at the time, just before the Sinai Campaign, was that the Flying School Qualified Flying Instructors were mostly ex-piston type fighter pilots that could not adjust to the new generation of fast jet fighters. Others were pilots that were inadequate to fly with the regular first line squadrons. As a consequence, the flying school was regarded by the fighter pilots as a place of exile in which promotion stopped. Sgan-Aluf Yesha'ayho Gazit accepted the command of the Flying School after the Sinai Campaign and became responsible for its major metamorphosis. The attitude towards the QFIs changed, and it was regulated that each fighter pilot, after spending a couple of years in a front line squadron, would be posted to the Flying School as a QFI as part of his regular promotion. To keep the QFIs in combat flying shape they continued to train and fly with their previous unit, once a week or more, as an emergency posting. Yet another change within the Flying Schhol was the shift to an all-jet syllabus. By the mid 1950s many types of basic jet trainers appeared in the west, and it was decided to select one of them for license production by Bedek Aviation. The types that were evaluated were the British Hunting Jet Provost and the two French jet trainers, the Fouga CM.170 Magister and the Morane Soulnier MS.769 Paris. Hugo Marom, IAI Chief Test Pilot, test flew all three types and recommended the Magister. By 18 December 1956 Aluf Tolkowski requested the initial procurement of eight Magisters, at a total cost of 1.2 million dollars, and in November 1957 the first Magister, registered 4X-FDA, arrived in Israel with Danni Shapira at the controls. However, more than two years had to elapse before the Magister entered IDFAF service since Bedek Aviation was busy organizing a production line for the French trainer. In February 1960 Hugo Marom made the maiden flight of the first aircraft ever produced in Israel when he took the first Magister, number 15, into the air from Lod International Airport to Tel-Nof, home of the air force Flying School. The era of the modern basic jet trainer had arrived...and, in fact, the Magister continues in the same role today, albeit much upgraded as the Tsukit.

Initially, two groups of cadets were formed: the bulk of them continued with the old syllabus while the smaller group was trained on the new jet. By the summer of 1962 at least 28 Magisters had been delivered to the Flying School and the old and faithful Harvards were finally retired. The Stearmans were also withdrawn from use, and the fighter pilot stream syllabus started with ten or less screening sorties on the Piper PA-11 Cub, followed by about 200 hours on the Magisters, which had an armament of two 7.62mm Browning machine guns and unguided 80mm air-to-ground rockets. A total of more than 100 Magisters were produced by Bedek Aviation before production ceased in 1966.

Thus, the Flying School changed dramatically during the late fifties and early sixties. Its servicemen, its aircraft and the syllabus were all changed, but yet another delicate point was not evaded by Aluf Weizman—the public image and the unique spirit of the air arm.

At first Aluf Weizman acted from within, and during 1959 the annual aircrew ball was introduced. Although he did not hold the patent rights to that event, since aircrew balls were held previously, too, but he turned it into a powerful tool to boost the morale of the IDFAF aircrews and to unite them. That event usually took place on Israel's Independence Day, and it was a re-union of all the aircrews within the IDFAF, with each unit responsible to create a short art show. Each year the aircrew ball was hosted by a different air base, and it contributed highly to the air force spirit, to its morale and to the brotherhood between the aircrews' themselves.

Although Weizman had turned his attention to areas, like the one above, that cannot be considered at first sight as the traditional missions of an air arm, he did not neglect the operational readiness

A pair of camouflaged Super Mystères of 105 Squadron, escorting an El Al Boeing 707.(GPO)

of the Chel Ha'avir as well, but the period of the late fifties and early sixties was relatively quiet with operational activity being restricted to casual skirmishes along the borders. Thus, the air arm was involved in restricted operational activity that consisted of three main types of roles: interception, reconnaissance and casualty evacuation.

The number of interceptions during that period was quite low, while the number of successful interceptions was even lower—much lower, in fact. The days in which Israeli fighter pilots achieved their astonishing kill rates in air-to-air combat were still ahead, and not a single IDFAF fighter was lost Several were damaged, and only a single Egyptian MiG-17 was officially recognized as a kill.

On 19 October 1958 a pair of Mystères tried, unsuccessfully, to intercept a Jordanian commercial Dakota that overflew the Ramon Crater in the Negev. The Jordanian pilot evaded successfully the two jet fighters and the Jordanians later claimed that the Mystères fired at the Dakota that contained several U.S. citizens, although the Israelis rejected that claim. Two months later, on Saturday, 20 December 1958, a pair of Mystères from Hatzor air base reported witnessing a pair of Egyptian MiG-17s flying at 30,000 ft over Nitzana. Sgan-Aluf Mordechai Hod, the Hatzor air base C.O., secured permission from Aluf Weizman to organize an ambush, and once the permission was granted, two pairs of Mystères were dispatched to the area. One of the pair intercepted two MiG-17s, and in the ensuing air combat one of the MiGs was downed near Bir Lahfan in Sinai and the second was damaged. However, that successful air combat was not the rule at the time, but rather the exception, and in the seven engagements that took place during the coming four years not a single Israeli pilot was credited with a kill, although the Egyptians lost a single MiG-17 due to an uncontrolled spin.

The first of those engagements took place on 27 April 1959 when a pair of Mystères intercepted without any results three Egyptian MiG-17s that penetrated into Israeli air space in the Negev. Next came the first combat appearance of the new Super Mystère in November 1959, but once again the engagement with the Egyptian MiGs was inconclusive. Following the Taufik incident, to be detailed later, the first air combat since the War of Independence between Israeli and Syrian fighters took place in February 1960 when Mystères and MiG-17s exchanged shots, without success to either side, with the Syrians falsely claiming one of the Israeli fighters. On 25 May 1960 a pair of Super Mystères intercepted four Egyptian MiG-17s over the Negev, and the Israeli pair claimed to have scored hits on one of the MiGs, but to the disappointment of the Israeli pilot the Egyptian regained his base successfully. Nevertheless, the gun camera film that captured the SMBD scoring hits over the MiG was proudly shown in movie halls all over the country for several months during that summer. Less than three months later, on 19 August 1960, another inconclusive engagement between Super Mystères and Egyptian MiG-17s took place. Finally, on 28 April 1961 a pair of Super Mystères were scrambled to intercept two Egyptian MiG-17s that overflew Nitzana and, just like in the previous engagements, an inconclusive fight developed, only this time one of the Egyptian pilots entered a spin and was seen to eject on the Egyptian side of the fence.

On 21 March 1962 yet another air combat occurred over the Syrian border but, as usual by this time, the pair of Super Mystères were unable to bring results, while this time things got even worse since one of the Syrian quartet of the MiG-17s slightly damaged one of the Super Mystères that successfully regained its base. Yet three other operational interceptions took place at the time, the first on 19 November 1959 when a pair of Mystères intercepted a Lebanese commercial Dragon Rapide that intruded Israeli air space near Nahariya and forced it to land at Haifa, the second, six months earlier, two 109 Sqn. Mystères flown by Segen Tsuk and Seren Yariv intercepted an old Lebanese AF SM.79MB transport that had

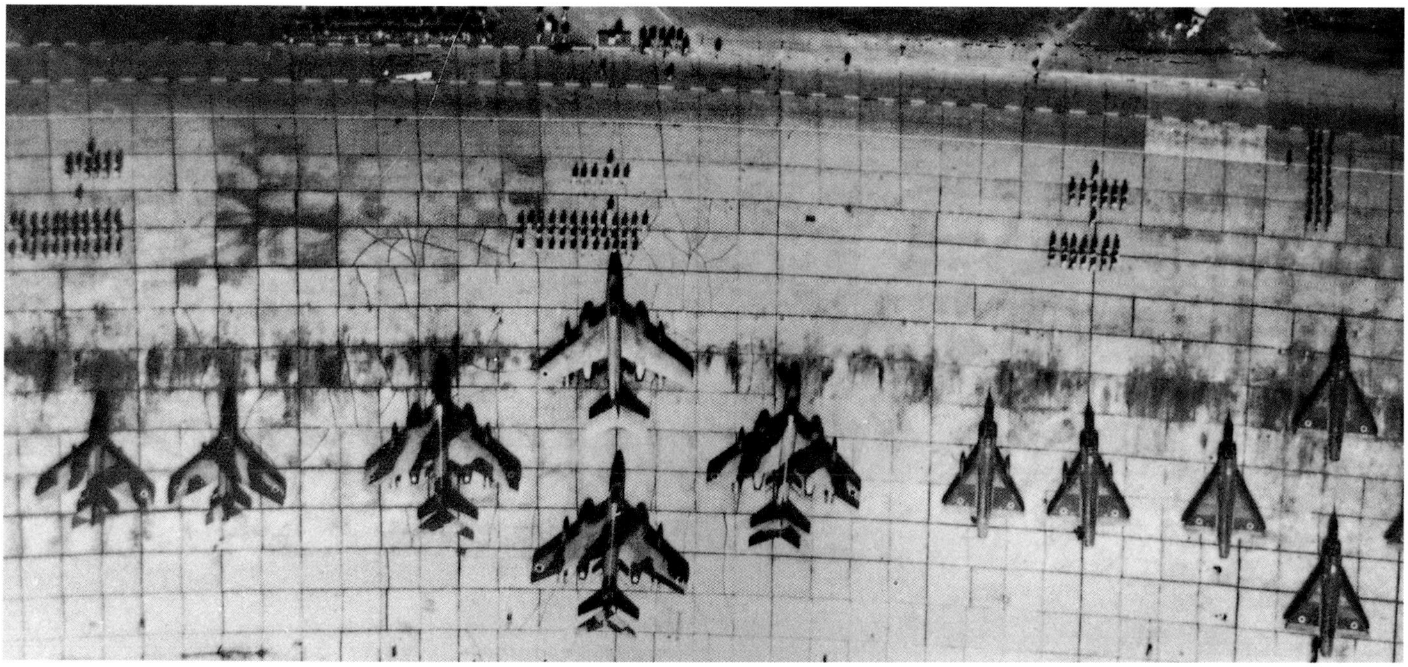

The sixties saw the incorporation into the Israeli Air Force of the powerful Dassault Mirage IIICJ and BJ, which eventully equipped 101, 117 and 119 Squadrons, providing, together with the long range Vatours, for the amazing victory over the Arab air forces during the Six Day War of June 1967.(IAF Magazine)

strayed into Israeli airspace, forcing it to also land at Haifa, and three days after that fighters from the same squadron forced a U.N. Dakota to land at Lod. To close the chapter of interception it must be mentioned that the U.S. showed particular interest at the time in Israeli nuclear activity and U-2 spyplanes were used to fly over the Negev to photograph the nuclear research reactor near Dimona. The efforts to intercept these flights were codenamed "Jule Verne", and at last on one occasion IDFAF Mystères and Super Mystères were scrambled to intercept, but obviously without any success.

IDFAF helicopters made their maiden casevac mission in combat on 17 March 1962 when the Golani Brigade attacked the Syrian strongpoint of Nokieb, east of Lake Kinneret (Sea of Gallilee), as a retaliation of the continuous bothering of Israeli fishermen. The attacking force suffered quite a large number of casualties, and these

After the arrival of the Super Mystères, the Mystère IVAs were quickly camouflaged, changing their main role to that of ground attack, as seen by these two jets from 101 Squadron. (IAF)

were evacuated by S-58s. Other Chel Ha'avir combat aircraft took part in that border skirmish when Syrian artillery opened fire at the Israeli forces, and the aircraft were ordered to counterattack the positions of the Syrian artillery. But, it was another IDF retaliation action on the Syrian border that culminated in a total intelligence failure on the Egyptian border. On the night of 31 January/1 February 1960 the Golani Brigade attacked the Syrian strongpoint of Taufik just southeast of Lake Kinneret. The Syrians replied with massive artillery fire, and in the coming days Israeli and Syrian fighters met in the air without any significant results. However, due to political considerations Nasser, Egypt's President, decided to ease the Israeli pressure on the Syrians by threatening its southern border with a huge concentration of forces in the Sinai Peninsula.

On 18 February 1960, the Egyptian Army 4th Armored Division, which was normally based west of the Suez Canal, commenced to cross the waterway eastwards into Sinai. This dramatic change in the capability of Egyptian forces in Sinai was completely missed by Chel Ha'modyin, the Intelligence Corps, and only an American hint caused Israel to take a closer look at Sinai. On 23 February a single Vatour IIBR, piloted by Seren Yosseph Sarig, a flight commander of 110 Squadron, was escorted on its reconnaissance mission along the Suez Canal by a single Vatour IIA flown by RAV-Seren Ya'acov Agasi, the commander of 110 Squadron. A quartet of Super Mystères was to remain on CAP over the Mediterranean Sea in case of trouble for the endurance of the sortie. The Vatour pair flew from Ramat-David air base, south to the Gulf of Aqaba, turned west to cross the southern part of Sinai, and then north to follow the Gulf of Suez and then the Suez Canal until turning east above the Mediterranean Sea. Four Egyptian MiG-17s were scrambled to intercept the Vatours, but failed to climb quick enough. The result of the risky sorties was definitively clear: the 4th Armored Division was no longer west of the Suez Canal. Thus, a second reconnaissance mission was dispatched the same day to Sinai and its results were amazing. The division was already at Sinai with between 400 to 500 tanks, versus the 20 to 30 Israeli tanks that were at the time in the Negev.

It was a major shock to the IDF General Staff, and on 24 May Aluf Itschak Rabin, Head of the Operations Department, claimed that until the IDF ground forces would not be redeployed to the Negev and the reservists mobilized the balance was made up of only the air force, nothing else. However, the IDF quickly concentrated its forces in the Negev, with only a partial mobilization of the reservists and Nasser lost his opportunity. By the end of the month the Egyptian forces returned to their barracks west of Suez and the show of force was over, although the feeling in Israel was that the Intelligence failed and the IDF was saved by a very small margin of luck and global policy.

The daytime reconnaissance executed by the Vatour IIBR during the February 1960 crisis was typical of the period. However, as the EAF brought to operational status its first units with its first level supersonic fighter, the MiG-19, it became clear that sending the Vatour IIBR out in daylight would become quite dangerous. The solution found was to use the night, since the Egyptians only had a handful of nightfighters, the Meteor NF.13 and a variant of the MiG-17. The first operational night recce mission was flown by the two flight commanders of 110 Squadron, Seren Yosseph Sarig and Seren Reuben Har'el, under the full moon of 23 January 1962.

The last issue of the period was the temporary disbanding of two squadrons. In January 1961 116 Squadron at Tel-Nof air base was disbanded and its Mustangs, the last combat piston type in IDFAF service, were phased out after more than a dozen years of faithful service. Several months later, in the summer of 1961, 116 Squadron reformed, again at Tel-Nof air base, as a Mystère IVA unit. The Mystères allocated to 116 Squadron were drawn from 101 Squadron, which was temporarily disbanded in that summer with the aim of getting air and ground crews ready to become the first Chel Ha'avir squadron to operate the new Mach 2 fighter, the Dassault Mirage IIICJ.

The single seat Vatour IIAs of 110 Squadron provided for many years the long range muscle for the Israeli Air Force. During the Six Day War they bombed Egyptian airfields as far as the Upper Nile in Egypt as well as Iraqi airfields in the west of that Arab country.(IAF Magazine)

Appendix: The British Air Attache's Secret Files

The documents presented herein contain the British Air Attache's annual report for the Israeli Air Force in 1957, just a year after the Suez War. It is very interesting reading, and includes the correspondence maintained with the Foreign Office HQ at London about the whereabouts of Israeli pilots in France! This shows an open secret that attaches the world over are closely related in intelligence gathering purposes, besides their "diplomatic" tasking.

....

No. 132
(1222/57)

British Embassy,
Tel Aviv.

December 9, 1957.

SECRET

V/A 1223/5

Sir,

I have the honour to transmit herewith a report on the Israel Air Force for the eleven months ending on the 30th of November, prepared by my Air Attaché, Wing Commander J.G. Bishop, D.F.C. As he will be leaving this post before the end of the year his successor will cover December in his annual report for 1958.

2. The report shows 1957 to have been a quiet year for the Israel Air Force; one dedicated to consolidation and training. While it has not this year received any new accretion of strength there is evidence that fifteen French Vautour aircraft will be delivered shortly. Whether or not they come it seems quite possible that attempts will be made before long to obtain a number of modern bombers as a deterrent to Arab attack on Israel's vulnerable cities. Her need for night fighters she has long since impressed upon us.

I have the honour to be,
with the highest respect,
Sir,
Your obedient Servant,

[signature]

The Right Honourable
Selwyn Lloyd, C.B.E., M.P.,
etc., etc., etc.,
Foreign Office, S.W.1.

SECRET

BRITISH EMBASSY,
PARIS.

? General.

September 30, 1957.

CD. 11014.

With the Compliments
of His Majesty's Embassy.

VR1223/4

Levant Department,
Foreign Office,
LONDON S.W.1.

JAR/JD.

CD. 11014.

British Embassy,
Paris.

September 30, 1957.

Dear Department,

Bordeaux reports that Air Fouga have announced that their light trainer Fouga 170 Magister is to be manufactured under licence by Bedek Aircraft Ltd in Israel.

2. We are copying this letter to Levant Department in the Foreign Office and to Tel Aviv.

Yours ever,

Commercial Department.

Commercial Relations and Exports Department,
Board of Trade.

SECRET

10612/68/57

BRITISH EMBASSY,

PARIS.

September 13, 1957.

WF 1224/3.

Dear Rose,

In his letter to Ross No. 10612/55/57 of July 11 the Ambassador referred to a report that 9 Israeli pilots had arrived at the French Air Force base at Dijon.

2. You may like to know that the Air Attaché had a further report from the same source that, on August 16, the 9 pilots were still at Dijon. They were all officers, the most senior being a Captain; they were using false names and had been issued with new identity cards by the Israeli Air Attaché in Paris. Although they had been flying Mystère 4 A's they had not been doing anything worth while with them or learning very much. The source was convinced that they were not at Dijon for this purpose.

3. A possible explanation of the presence of these officers in France has occurred to the Air Attaché. He thinks that they may have come here to ferry back to Israel the ground attack version of the Vautour. About 15 of these have recently been built at St. Nazaire, but it seems that the French Air Force already have no use for them. They might however be extremely useful to the Israelis, for instance in taking action against the forts guarding the Straits of Aqaba. A piece of supporting evidence is that one of the officers (a Lieutenant of German extraction) left for Tours, where the Vautour had started to arrive from the St. Nazaire factory, about the middle of August.

4. This is of course only a theory, but it seems sufficiently plausible and interesting to bring to your notice. The Air Attaché has already reported direct to the Air Ministry.

Yours sincerely,

Tony Duff

(A. A. Duff)

E. M. Rose, Esq., C.M.G.,
Levant Department,
Foreign Office, S.W.1.

(VJ 1223/15) FOREIGN OFFICE, S.W.1.

CONFIDENTIAL May 2, 1957

You wrote to me on April 9 and 11 (your reference BJ.112/5/S.6) about the question of attendance by Jordanians and Israelis at Royal Air Force courses in the United Kingdom.

2. As regards Jordanians we should like to defer our final view until we have had a chance of seeing where present developments are taking us and of determining our general future policy after discussions with the Americans. On the face of it there would seem to be something to be said for trying to keep alive the association which has grown up between the Royal Air Force and the Royal Jordan Air Force and thus sustain some measure of British and Western influence in Jordan.

3. We feel that the Israelis should be neither denied places altogether nor offered special facilities but should take their chance along with all other applicants. In deciding which applicants to accept on any particular course I assume that you would normally take into account the relevance of the course to the degree of efficiency and the type of equipment used by the applicant's Air Force, and that if, for example, there is a toss-up between two applicants for one vacancy at a Hunter course, the man who is more likely to fly Hunters is presumably the one to take.

4. All the above is, of course, on the assumption that there is no strong security reason against accepting a Jordanian or Israeli on any particular course.

(P.C.H. Holmer)

Miss M. McClelland,
Air Ministry,
S.6.

From: Miss M.McClelland

Tel. N[illegible]
TRAF[illegible]AR 8811, Ext. 6331

AIR MINISTRY,
WHITEHALL GARDENS,
LONDON, S.W.I.

BJ 112/5/S.6.

R1223/2

11th April, 1957

Dear Mr Holmer,

When I wrote to you yesterday seeking advice on the attendance of Jordanians and Israelis on R.A.F. courses in the U.K, I said that the Israeli Air Force had not asked for any places since the ~~Exercises.~~ Suez crisis.

This morning, however, the foreign liaison staff here received a letter from the Israeli Air Attaché in London asking for places on a variety of courses in 1957/58. Your advice will therefore be of more than academic interest. One of the courses specified involves flying on Hunter aircraft. If room could be found there for an Israeli student we should have to consider the wisdom of encouraging the Israeli Air Force to become familiar with a type of aircraft which they might still be prevented from acquiring.

Yours sincerely,

M. McClelland

P.C. Holmer, Esq.,
Foreign Office,
S.W.1.

S E C R E T.

Appendix "E" to
AA.TEL,'S.83/Int dated
1st. December,1957.

OFFICIAL VISITS BY AIR ATTACHE
JANUARY - NOVEMBER,1957.

Below is a list of the official ceremonies and visits which the Air Attache attended during the period covered by this report. The Air Attache was invited to participate in twenty three other visits, but as these were with one exception of a purely army or naval nature they were declined with thanks.

23rd. January	-	The A.O.P. Piper Cub Squadron at Ramle.
1st - 4th. April	-	Civil Air Attache, Beirut.
29th. April	-	I.D.F. Supply Corps.
5th. May	-	Memorial Day, Israel Air Force Dead.
6th. May	-	Independence Day Parade and Presidential Reception.
13th. May	-	Israel Air Force Station Ramat David.
27th. - 30th. May.	-	Headquarters Middle East Air Force and Air Headquarters Levant, Cyprus.
26th. August - 5th. September.	-	Assistant Chief of Air Staff (Intelligence)'s Air Attaches' Conference, The Air Ministry.
21st. October	-	Armoured Corps Exercise.
22nd. October	-	I.D.F. Signals School.

suggest we send a copy to Washington and relevant extracts to Paris.

[signature] 7/1

I think it would be useful to have a note made on the basis of these extracts for submission (with appropriate Arab comparisons)

[signature]

[initials]

Circulate to P.U.S.D.
E.R.D.
I.P.D.
G.R.D.
News D

Number of copies ....

Number of copies of enclosures ....

1957	V LEVANT DEPARTMENT	VR1223/5

FROM Sir F. Rundall, Tel Aviv.

SECRET.

No. 132 (1222/57).

Dated December 9

Received in Registry— December 17.

Israel Air Force January 1st November 30th 1957.

Report by Wing Commander Bishop, Air Attaché.

References to former relevant papers

(Print)

(How disposed of)

(Action completed)	(Index)
EMR 3/4	[illegible] 25-11-58

References to later relevant papers

M4668 51893—1

MINUTES

~~Copy~~

Air Ministry & Min. of Defence receive this automatically.

2. There is little new here but it is worth looking at paras. 7-9 on French influence; 26-29 on Israel's increasing self-sufficiency, or efforts towards it; and 30-32 on Israeli attitude to the Armistice Agreement when it comes to infringing Arab air space. Quality remains good.

[illegible] 7.1

Miss [illegible] 7.1

E.R.D. (Mr. Doyle)

P.U.S.D.

I think you may find the paras mentioned by Mr. [illegible] interesting. Also para 20 which has relevance when people talk about a "balance" of arms. Operative Jordan jet pilots can be counted on the fingers of one hand and the Iraqis probably only have 10-15. Para 29 is of great future importance.

After action

/ [illegible]

V

LEVANT DEPARTMENT

ISRAEL

VR1223/4

1957

FROM
Commercial Dept.,
Paris
(Comm'd).
No. CD 11014
Dated September 30
Received in Registry— October 10.

Air Fouga have announced that their light trainer Fouga 170 Magister is to be manufactured in Israel.

References to former relevant papers

MINUTES

Mr. Unwin PMI 14/10.

ERD (Mr. Doyle)

15/10

16/10

(Print)

(How disposed of)

(Action completed)
16.10.57.

(Index)
25 Oct. 58

References to later relevant papers

49847

1957	V LEVANT DEPARTMENT ISRAEL.	VR 1223/3

FROM SECRET
A. A. Duff,
Paris,
to Mr Rose.
No. 10612/68/57.
Dated September 13
Received in Registry— September 17.

Nine Israeli pilots are still at the French air base at Dijon, using false names and carrying indentity cards issued by the Israeli Air Attaché in Paris.

References to former relevant papers

VF 1224/3.

MINUTES

Western Dept.

Pus D.

Eastern Dept.

(A) Air Ministry (Communicated) AA/S.1220/3/Int. 19/9/57. (initialled within)

(Print)

(How disposed of)

(Action completed)

(Index)

References to later relevant papers

49724

V

LEVANT DEPARTMENT

VR 1223/2

FROM
Miss M. McClelland
Air Ministry to
MR Holmer.

No. BJ 112/5/S.6.

Dated April 11

Received in Registry— May 1.

Training courses in the U.K.

Has received a request from the Israeli Air Attaché for places for Israeli students on various courses during 1957/58, including one on flying Hunter aircraft.

References to former relevant papers

VJ 1223/15.

MINUTES

See minutes at VJ 1223/15.

3/5

(Print)

(How disposed of)

Dft. Miss McClelland, Air Min.
from Mr. Holmer. May 2.
(desp'd on VJ1223/15)

(Action completed) 3/5

(Index) 25.11.58

References to later relevant papers

49847

AA/S.1220/3/Int. **SECRET** 732 /57

19th September, 1957

R1223/3(A)) PW

Intelligence Report No. 222/57

Israeli Air Force

Attached is a copy of a letter sent by Chancery to the Foreign Office on the subject of the Israeli pilots and Air Traffic Controller at present serving with the French Air Force.

2. An A.2 source reported to the Air Attaché to-day that 8 Israeli Air Force Lieutenants, (i.e. 4 crews of 2,) are at present at TOURS. 2 Crews are undertaking night fighter training on the Vautour and the other 2 crews are doing likewise on Meteor NF. 11s.

3. It is rumoured on the base that the Israelis will eventually take delivery of some Vautours.

4. Chancery would be grateful if the above information could be passed to the Foreign Office to be read in connection with Chancery letter 10612/68/57 dated 13th September, 1957.

(C. E. S. LOCKETT)
Air Commodore
H. M. Air Attaché
PARIS

WR1223/3

PWU 27/9

DDI 2

AA Tel Aviv

A 1746/2

Copy to : Chancery. **SECRET**

E. M. Rose Esq. C.M.G.
Levant Department
Foreign Office
S.W.1

With the Compliments of

The Under Secretary of State

for Air

A. C. 2. (a)

Whitehall Gardens
S.W.1.

S E C R E T.

AA.TEL/S.83/Int.

Office of the Air Attache
British Embassy,
Tel Aviv.

SECRET

1st. December, 1957.

Sir,

I have the honour to submit a report on the Israel Air Force for the period 1st. January to 30th. November, 1957. I expect to leave this post during December and it would be unfair to expect my successor to write the annual report for 1957. December of this year had best, therefore, be included in the annual report for 1958.

INTRODUCTION.

2. Last year was notably violent, culminating in the Sinai Campaign. This year has been unusually calm. Whereas in 1956, up to the time of the Sinai Campaign, there had been nearly three hundred Israeli casualties on the frontier, including nearly eighty dead, this year to date only twenty three Israelis have been killed in border incidents. And of these deaths eight could not be put down to Arab mayhem as six resulted from foolhardy Israeli excursions deep in Arab territory and two were caused by Israelis driving over their own mines. The majority of incidents have been of a very minor nature and have involved thieves, smugglers and Arabs endeavouring to cross the Negev between Egypt and Jordan.

3. This year has been one of thought and consolidation rather than one of activity and innovation for the Israel Air Force. Much time has been spent thinking over and digesting the lessons learnt from the Sinai Campaign. Though flying has continued at a reasonably high intensity there appears to have been a definite slackening from the high pressure training which took place during 1956 prior to Operation Kadesh.

4. For ease of reading I have placed details of the Israel Air Force in Appendices as follows:-

Appendix "A" - Summary of aircraft in Israel as at 30th. November, 1957.
Appendix "B" - Israel Air Force - Order of Battle as at 30th. November, 1957.
Appendix "C" - Israel Air Force - Organisation at Station Level.
Appendix "D" - Senior Appointments in the Israel Air Force as at 30th. November, 1957.
Appendix "E" - Official Visits by the Air Attache during 1957.

ISRAEL AIR FORCE.

POLICY.

5. The policy for the Israel Defence Force still requires the armed services to be able to fight and defeat the combined Arab States. Further the prime importance of winning the air battle is fully appreciated. This was made clear by the Director General of the Ministry of Defence in May when he said, "The threat to Israel has not yet passed and it is up to us to make a new and intense effort to prepare for any eventuality. The Arab States possess the most up to date arms and we have to provide the Israel Defence Force with the necessary weapons, particularly those used in the air, to enable us to prevent war in the future." However, Israel realises it is impossible to purchase and maintain everything desirable in the way of weapons, especially air weapons. This applies particularly in the maritime role now the Arab Navies have increased so much in size and acquired submarines. This was reflected in a remark of the Commander-in-Chief, Aluf Tolkowsky, when in July he said to your Air Attache, "Russian submarines in Egyptian hands are a question of Russia versus the United States. We shall notbe rushing to you asking for Gannets or Neptunes!"

/ para 6.

His Excellency Sir F[illegible] Rundall, K.C.M.G., O.B.E.,

SECRET.

- 2 -

6. If it is not practical for the Israel Air Force to build up coastal forces, and her fighter/ground attack force is not unsatisfactory, there are nevertheless two sections of her air force so dangerously in need ofre-arming they have virtually ceased to exist. These are the night fighter element and the bomber force. Both roles were previously filled by Mosquitos. The necessity for night defence is self-evident. It is worth repeating, however, that Israel has always relied for the safety of her three large towns - Tel Aviv, Haifa and Jerusalem - on the Arabs knowing that one bomb on an Israel city will bring heavy retribution in Cairo, Damascus and Amman. There has arisen therefore a strong feeling that modern bombers must be acquired, and if aircraft capable of tactical operations and reconnaissance in addition to fulfilling the bomber role are available, such machines will be the choice of Israel.

AIRCRAFT.

7. There has been no evidence that the Israel Air Force has taken delivery of additional aircraft so far this year, but it is believed that they have purchased some fifteen Vautour twin jet aircraft from France. Whether these aircraft will be the bomber/ground attack version or night fighters is not known definitely, but it is believed they will be bombers and if so they will be a potent weapon in the Middle East. Though these aircraft have not yet been delivered one has been seen in France bearing Israel national markings, so delivery cannot be far off.

8. Israel has investigated the performance, so far as permitted or possible, of most of the aircraft produced in the western world that could possibly be ofuse to her armed forces. In particular, interest has been shown in helicopters. Two Italian Agusta built Bell light helicopters were in Israel during the autumn for demonstrations, as also were one or more French Alouettes. There is however no news of a contract, but a purchase of one or other type seems highly likely. It is stated on fairly good authority that an order has been placed in America for three Sikorski S.58 helicopters. These are larger than the S.55 model which Israel already possesses, carrying up to sixteen men, and can be used in the normal variety of helicopter roles. They have probably been ordered in lieu of three S.55's which were not delivered because of the arms embargo.

9. The other main item of interest is the conclusion of a contract between the Bedek Maintenance Base and the French Firm of Air Fouga for the construction of Magister jet trainers at Lod. This programme is still in the planning stage, but it seems certain that the Flying Training School will eventually re-equip with this type, though hardly before late 1958. This machine can carry rockets and guns, which combined with the Israeli's rather care free way of hanging on more than the text book advocates could turn the Magister into a useful close support operational aircraft.

10. It seems probable that the accident rate in the Air Force is about comparable with that of the Royal Air Force. In November therefore the figures for aircraft holdings were written down appropriately. The revised assessments are given at Appendix "A" to this report.

11. Israel has had almost no success in disposing of the large collection of obsolete aircraft which litter the hardstandings at Lod, Kfar Sirkin and Ekron. One Catalina was sold in the United States, and the first of three C.46 Commando transports was also delivered to America. The other Commandos are still at Bedek as the purchaser is allegedly short of money.

ORDER OF BATTLE.

12. These aircraft form an order of battle but little changed from that given a year ago in the Annual Report for 1956. It is obvious from my remarks above that the strength is virtually the same, and there does not appear to have been any major or permanent re-deployment.

/para. 13.

SECRET.

- 3 -

13. The air force is still basically a small tactical group consisting of four jet fighter/ground attack squadrons with small supporting transport, reconaissance and night fighter units. In addition there is an efficient and wide spread A.O.P. element deployed with the Army and armed with Piper Cub aircraft. The whole based on reasonable flying and technical training establishments.

14. As with larger western air forces the Israelis have found maintaining a reserve in the age of swept wing fighters extremely difficult. As the piston engined aircraft become unbattleworthy it is found impossible on economic and technical grounds to replace them with jets. So whereas before the Sinai Campaign they could muster some three squadrons of fighter/ground attack aircraft mannedby reservists, this second line force has now dwindled to the equivalent of one Mustang Squadron. Early in the year the Mosquitos and B.17's were still being kept serviceable as a rather amorphous reserve, but it seems likely this technical effort has been dropped or at least fallen to a very low level. The acquisition of the most modern aircraft and anti aircraft guns by Israel's deadliest enemies is in no small measure responsible for this change of policy.

15. In the Order of Battle at Appendix "B" to this report the dispersed Piper Cub A.O.P. flights which come under Army Command are shown separately, last year No. 100 Squadron was shown as a consolidated figure at Ramle. The RCM B.17 has been dropped as it has been at Bedek for many months. The only other changes concern the reserve.

AIRFIELDS.

16. There is little to say on the subject of airfields. With its modern jet aircraft the Israel Air Force is tied to three main bases - Ramat David, Ekron and Hatsor. Even Ekron is small for the more advanced aircraft having a maximum run of 2,000 yards. Lod is the only other airfield large enough to be used in an emergency. The secondary abandoned airfields have continued to deteriorate slowly and time and money do not appear to have been wasted on them. The one exception apart from airfields used by civil aviation is Kfar Sirkin. This is a good but small airfield and could once again be a home for the Flying Training School should Ekron become overcrowded.

17. There was evidence of some new building at Ramat David during the autumn. Its nature is not yet clear but possibly it involves the preparation of offices and technical buildings, including a hangar, for the expected Vautour squadron. Minor works have also been carried out at Lod but they are better recorded in my report on Civil Aviation.

CONTROL AND ORGANISATION.

18. There has been no change in the overall organisation of the Air Force. The Air Officer Commanding is still both Commander-in-Chief and Head of the Air Staff. The base commanders are still directly responsible to the A.O.C. In other words the Air Force remains one small group without any "Command"structure.

19. There has been a slight change in the form at station level. The traditional three wing system has given way to squadron commanders being directly responsible to their base commander. This was found desirable because of the diversified roles of the flying units, making it almost impossible for an Officer Commanding Flying Wing to be as experienced as all his squadron commanders. This re-organisation increased the span of control of the base commander to straining point. Consequently the post of Deputy Base Commander was introduced. At Appendix "C" is shown the present organisation of a station.

/FLYING PERSONNEL.

SECRET.

- 4 -

FLYING PERSONNEL.

20. In January, 1957, it was estimated that there were 194 pilots on regular service with the Israel Air Force. Since then there must have been at least two wings parades. The total number of pilots today is probably therefore between 200 and 220. This figure includes staff officers but they are mostly in training on modern aircraft. These figures do not however include A.O.P. pilots who wear "light aircraft only" brevets.

21. Last year I suggested that unless pay and promotion prospects were improved the air force would find it difficult to get good quality young men as officers. During 1956 there was a considerable increase in flying pay and the "glory" the air force acquired during the Sinai Campaign was also a great help with recruiting. There have been other financial palliatives during 1957, for example officers no longer have to pay for meals eaten in camp, senior officers receive fixed expense allowances without having to submit bills and all officers are to be issued with a free dress uniform. However, the ultimate problem remains, the lack of prospect of a reasonable career. The chances of a pupil pilot becoming a Sgan Aluf (Wing Commander) are pretty slim. There are only five Aluf Mishne (Group Captain) posts for flying personnel in the service and they very rarely fall vacant as the incumbents are still relatively young men.

FLYING TRAINING.

22. Evidence suggests that flying training has continued on sound lines. According to Tolkowsky the standards required of prospective cadets have been raised during the year. Pupils go through a grading process on Stearman bi-planes which they fly for not more than thirty hours. They then do the main part of their training on Harvards, and finally fly about fifty or sixty hours of jet conversion on Meteors before receiving their Wings. The Meteor training includes about ten hours night flying. After graduation they are posted to the Ouragan Squadron which serves as an operational training unit. It must be assumed that the products of this training are today the equivalent, as Fighter/Ground Attack pilots, of the products of our own air force.

GROUND CREWS AND GROUND TRAINING.

23. The Technical Training School at Haifa continues to turn out tradesmen of a fairly reasonable standard though because of the relatively short length of the course the products are doubtless less skilled and knowledgeable than those of Halton or St. Athan. But because of the relatively simple equipment of the Israel Air Force and the operational conditions of the Middle East, the ground crews are not unsatisfactory. Among more senior tradesmen there is plenty of experience and ability.

24. The Israel Air Force Staff School continues to function at Ekron and it appears to give good and valuable background to junior staff officers. Graduation from this course is still a necessary qualification for promotion to Rav Seren (Squadron Leader). It was announced during the year that the Army and Air Force were establishing a special school of Land/Air Warfare. This was in fact a revival of an establishment which was closed some two years ago. It is a small edition of our own establishment at Old Sarum.

25. Much money and effort has been put into Air Gadna and senior officers always seem convinced that this effort with youth channels the best quality and most enthusiastic conscripts to the Air Force.

/MAINTENANCE.

S E C R E T.

- 5 -

MAINTENANCE.

26. Major servicing and repair remain centralised in the Central Maintenance Unit at Ekron and at Bedek, Lod. Perhaps the most important development in the Israel Air Force this year has been the ability of Bedek to organise and equip itself for the repair and inspection of jet aircraft and engines despite the difficulties they had in obtaining tools and material as a result of the arms embargo following Sinai. They are now servicing Nenes and Derwents and have probably started work on Verdons, the power plant of the Mystere. There is no doubt that this maintenance base can undertake almost any aeronautical work and do it well.

LOGISTICS.

27. Logistics continue on a common-user basis which means the Air Headquarters staff is almost entirely concerned with technical equipment problems. It is doubtful if there have been any problems due to shortage of spares thanks to the French friends. There may have been a tight period in Derwent engine spares until the British Government decided to release Rolls Royce equipment in July and August, but this did not appear to affect seriously flying training. One cannot but wonder if France was not able to help with Derwent maintenance as there are still Meteor Night Fighters in the French Air Force.

28. Experience of Foreign Governments turning off the tap is undoubtedly a large factor in the Israeli desire to build up a native aeronautical industry. Bedek is the obvious nucleus of such an organisation and the intention to build Fougas there the first step in such a development. The Technion at Haifa has given almost top priority to producing aeronautical engineers and aviation research facilities and the Weizman Institute at Rehovoth has been undertaking research for the United States Air Force. Both the Tadir and American Eastern Electronics companies went bankrupt during the year. They produced valves, crystals, radar and communications equipment. It seems probable however that the government will not allow these valuable defence assets to be lost and they will probably be taken over by the Histadrut. Military industry has increased greatly during the year and Israel now manufactures 68 and 105 mm. Brandt rockets making them virtually independent of foreign sources for weapons and lethal stores.

29. From now on therefore arms embargos will be less and less effective so far as Israel is concerned. Naturally raw materials and petroleum products will always have to be imported, but the descendants of Joseph are unlikely to forget there are lean kine as well as fat beasts. Certainly there are already vast stocks of all things which are imported and vital for defence.

AIR OPERATIONS.

30. The four traditional types of air operation have been repeated this year but on a more minor scale. Photographic reconnaissance has been carried out with increasing caution as the Syrian and Egyptian Air Forces have been re-armed. The Mosquito is still used when a long range penetration is considered essential. Shorter sorties have been flown with jets modified to carry vertical cameras, including at least one Mystere. The Israelis also possess Dakotas modified for vertical photography, though it is doubtful if these have any practical value at the moment.

31. Tactically Piper Cubs are used for surveying of border areas, and they certainly infringe Jordanian air space and have probably also been used over Sinai. These machines have proved of great value against armed infiltrators and smugglers in the Negev, and their constant patrols have undoubtedly produced a high degree of safety for travellers in the southern desert areas.

/para 32.

SECRET.

32. No Arab aircraft have been shot down over Israel this year. The Israel Air Force claimed a Jordanian airliner crossed into Israel near the Wadi Ramon on 19th. October and they endeavoured to capture it. It seems certain however from many independent witnesses that the Israel fighters carried out their interceptions over Jordan and possibly also over Saudian territory. It has never been ascertained why this happened. The most obvious and likely explanation is over enthusiasm on the part of the pilots concerned.

33. Lastly the Transport Squadron has continued to fly to Western Europe, mostly to France via Italy, which has kept crews practised in those bad, weather conditions not available in Israel and provided good navigational experience.

RADAR.

34. The operational radar in Israel is still based on the traditional sites on Mount Carmel and Hatsor. What, if any, new equipment the Israelis have received since the introduction of the French "STATION RADIA" and "RAMOLO" sets during the summer of 1956 is not known. Some new apparatus has been erected on Mount Carmel but this has not yet been seen without a canvas cover on so its nature is not known. It is however accompanied by some new and extremely complex aerials. The seaward facing marine scanners are still in position at Stella Maris (Haifa) and Jaffa. There is no reason to alter the previous assessment that aircraft can be picked up at distances up to 160 miles from Hatsor and Carmel and up to altitudes of 60,000 feet.

RADIO COMMUNICATIONS.

35. There is little faith in lindlines in Israel, and with good reason. Apart from the physical ease with which such links can be broken or tapped, nowhere in Israel is far from the border. There has arisen in consequence a complex series of V.H.F. nets covering the country, and now additional U.H.F. links are being installed. Apart from the armed services the main organisations using V.H.F. nets are the Post Office, the Police and the Electricity and Water authorities.

36. There are more amateur radio operators per capita in Israel than in any other country in the world and many settlements have radio communications. Civil Defence is linked similarly. Communications in Israel in an emergency are therefore unlikely to fail catastrophically.

MORALE.

37. Morale remains high, though with the increase of Russian influence in Syria and Egypt there seems to have been a slight deflation of the extreme cockiness of former days. There is no reason to suppose that in the event of being attacked the armed forces of Israel would not fight desperately to the bitter end endeavouring on every possible occasion to act offensively.

38. The great influx of immigrants during the past year from Eastern Europe and North Africa has once again put back the movement to homogeneity in the population. One cannot say the civil population would react with one spirit in the event of war, the fortitude would vary from locality to locality depending on the make up of the local citizens.

ANTI AIRCRAFT UNITS AND CIVIL DEFENCE.

39 Israel is well supplied with light anti aircraft guns of may types and all vital points such as military establishments, bridges and ports are liable to put up fire against low flying aircraft. It is doubtful whether heavy anti aircraft batteries are taken very seriously today. They are kept in being more for the effect they would have on civilian morale than anything else.

/para 40.

S E C R E T.

-7-

40. Civil Defence (HAGA) continues to be active and in the past year its organisation and command system have been slowly but surely improving. As with all civil defence organisations one would expect that until Haga had been schooled in war its performance would be a little chaotic.

SECURITY.

41. There has been no change in Israel security policy vis-a-vis the service attaches. The exception is the Frenchman. There is no doubt that Israel loses much in goodwill while gaining little in security by the absurd lengths to which she carries her terrorism begotten secrecy and ill manners. No restrictions however have been put on the movements of the Service attaches.

ACTIVITIES. AIR ATTACHE.

42. A list of the official ceremonies, visits and talks which the Air Attache attended is given at Appendix "E". The number of air force visits was very poor this year and there were only two of a genuinely air nature. I have made some rather rude comments about this recently but have not pressed very hard for visits during the last two months, as it is better for them to take place as soon aspossible after I leave rather than have the Israelis able to turn to my successor and say - "we took Wing Commander Bishop there just before you arrived".

43. A visit was made to Headquarters Middle East Air Force in May for consultations with the Chief Intelligence Officer. The Civil Air Attache in Beirut was visited in April. I attended the Annual Conference of Air Attaches held by the Assistant Chief of Air Staff (Intelligence) in London during August.

44. The Israel branches of the Air Force Association are still in being and after a rather slow summer are increasing their activities. There is of course no new source of recruitment to these branches and as the average age increases the numbers inevitably dwindle.

FINANCE.

45. One can only say that the amount spent on defence is far from clear, and even if the actual allocation for the financial year was known it would probably mean little because the large Israel purchases in France were doubtless on a credit basis. In last year's report I estimated the 1957/58 Defence budget at about I.L. 222 million. This figure was based on I.L. 80 million for defence in the normal budget, I.L. 68 million in the form of a Security and Special Budget, and approximately I.L. 75 million in the form of Defence Levies. In addition there would be the cast of the Border Police.

46. When the Budget was presented in April the Minister said that I.L. 218 million in taxes and levies had been spent on security in the last eighteen months. This was certainly a gross understatement. There is no reason to assume that this year's total, including paying off debts, was any less than last year, so one must assume a figure of over I.L. 220 million.

APPOINTMENTS.

47. There have been few changes in high appointments. Aluf Tolkowsky is still Air Officer Commanding and Aluf Mishna Weizman heads the Air Division. A list of the more senior posts is given at Appendix "D".

/CONCLUSION.

S E C R E T.

- 8 -

CONCLUSION.

48. After the previous two years writing this report has been a rather miserable affair and it inevitably lacks interest. Most of your Air Attache's conversations and travels this year have resulted in a report of "no change". There have been no additions to the Air Force, no movements of units, no major changes in organisation, training or morale. The Air Force remains small but strong. A trim affair with its weight in the blade rather than in the hilt. An Air Force worthy of respect.

I have the honour to be,
Sir,
Your obedient Servant,

(J.G. BISHOP)
Wing Commander,
Air Attache,
Tel Aviv.

SECRET.

Appendix "D" to
AA.TEL/S.83/INT. dated
1st. December, 1957.

ISRAEL AIR FORCE

SENIOR APPOINTMENTS - 30TH. NOVEMBER, 1957.

NOTE.
(i) Previous names in brackets where these have been changed recently.
(ii) Asterisks indicate still in post within last month.

1. Air Officer Commanding - Aluf Dan TOLKOWSKY +

The Air Headquarters - Ramle

2. Chief of Staff - (vacant)

3. Air Inspector - Rav Seren Y. ARBEL +

4. The Air Division

Head of Division - Aluf Mishne A. WEIZMAN +

Training - Sgan Aluf Z. KENNY +
Rav Seren YAVNEH +

Intelligence - Sgan Aluf I. SHANA'AN (TESSEL) +
Rav Seren LERON (LONDNER) +

Operations - Rav Seren M. PELED +

Planning - Rav Seren ALDER +

Org. & Staff Duties - Sgan Aluf HERON +

5. Manpower Division

Head of Division - Aluf Mishne A. MEDAN +

Personal Services - Sgan Aluf ORON +

Manpower Org. - Sgan Aluf GEY-RON
Airmen Promotion & Remustering - Rav Seren E. BEN HORIN
Manning - Rav Seren FAST
Officers - Rav Seren RATZINA

Senior Legal Officer - Sgan Aluf JERUSHALMY+

6. Materiel Division

Head of Division - Aluf Mishne Y. RABIN

Quartermaster Stores - Sgan Aluf SHOHAM +

Air Armament - Rav Seren KESARLA +

Engineering Development - Sgan Aluf I. LAPID @

Maintenance Supply - Sgan Aluf POSSNER

Maintenance Engineering - Sgan Aluf HALPERIN @

Supply Planning - Rav Seren BAZAR @

(@ These three appointments may be mixed)

/7.

SECRET.

Appendix "D" to
AA.TEL/S.83/Int dated
1st, December, 1957.

- 2 -

7.	Chief Electronics Officer	-	Rav Seren ARNON +
8.	Principal Medical Officer & O.C. Medical Boards.	-	Sgan Aluf LAVI +
9.	O.C. 100 Squadron	-	Rav Seren M. KENER +

STATIONS.

10. Ramat David.

Commander	-	Aluf Mishne G. ELROM +
Deputy Base Commander	-	Rav Seren M. BAR +
Chief Technical Officer	-	Rav Seren ROTH
O.C. 117 Squadron	-	Rav Seren A. YOELI +
O.C. 109 Squadron	-	Seren SHEFER +
O.C. 119 Squadron	-	Rav Seren TSIDDON +

11. Hatsor.

Commander	-	Sgan Aluf M. HOD (FINE) +
Chief Technical Officer	-	
Senior Admin. Officer	-	Rav Seren YANAI
O.C. 101 Squadron	-	Rav Seren B. PELED +
O.C. 113 Squadron	-	Seren Pachik

12. Ekron.

Commander	-	Aluf Mishne LAHAT +
Deputy Base Commander	-	Sgan Aluf TURIN +
O.C. Flying Training School	-	Rav Seren GATZIT +
Chief Flying Instructor Flying Training School.	-	Seren SHEFER
Central Maintenance Unit	-	Rav Seren ALONI
O.C. 103 Squadron	-	Rav Seren H. ALPERSTEIN +
O.C. 115 Squadron	-	Rav Seren HASHILONI
O.C. Reserve Mustang Squadron	-	
O.C. I.A.F. O.C.T.U.	-	Sgan Aluf WASSERMAN
O.C. I.A.F. Staff School	-	Aluf Mishne G. SCHOCHAT

/13.

S E C R E T.

- 3 -

Appendix "D" to AA.TEL/S.83/INT. dated 1st. December, 1957.

13. Haifa.

O.C. Technical Training School	-	Sgan Aluf U. PERES (PROESS)
Chief Instructor	-	Rav Seren BARSELLA
Senior Admin Officer	-	Rav Seren A. LEVI

MISCELLANEOUS APPOINTMENTS.

14. Air Force Instructor at Army Staff College.

- Sgan Aluf SHATKAI

15. Foreign Liaison Officer.

- Rav Seren C. ZEVADI +

16. Air Force Representatives in Ministry of Defence and G.H.Q.

Manpower Branch	-	Aluf Mishne AYAL.
Purchasing Branch	-	Aluf Mishne B. COHEN +

17. Air Gadna

Staff Officer i/c Air Gadna	-	Sgan Aluf DVIR +
O.C. Air Gadna Training Base Tel Aviv.	-	Rav Seren BEN-OR.

18. Israel Air Attaches

Washington	-	Sgan Aluf A. ZIELONI +
Paris	-	Sgan Aluf P. KEDAR +
London	-	Aluf Mishne Y. NEEMAN (Army Officer)
Air Force Representative on Israel Purchasing Mission in Paris.	-	Sgan Aluf Y. GILADI

19. O.C. Equipment Depot Sarafand. - Sgan Aluf H. GEYOR.

20. O.C. Intelligence Unit, Ramle - Sgan Aluf KRIS +

21. LOD Airport.

Liaison Officer with Bedek - Tav Seren ISH- SHALOM +

22. On Leave for study purposes.

- Sgan Aluf A. YOFFE +

- Sgan Aluf B. KERET +

Appendix "C" to
A.A.Tel/S.83/Int dated
1st. December, 1957.

ISRAEL AIR FORCE

ORGANISATION OF A STATION OR BASE.

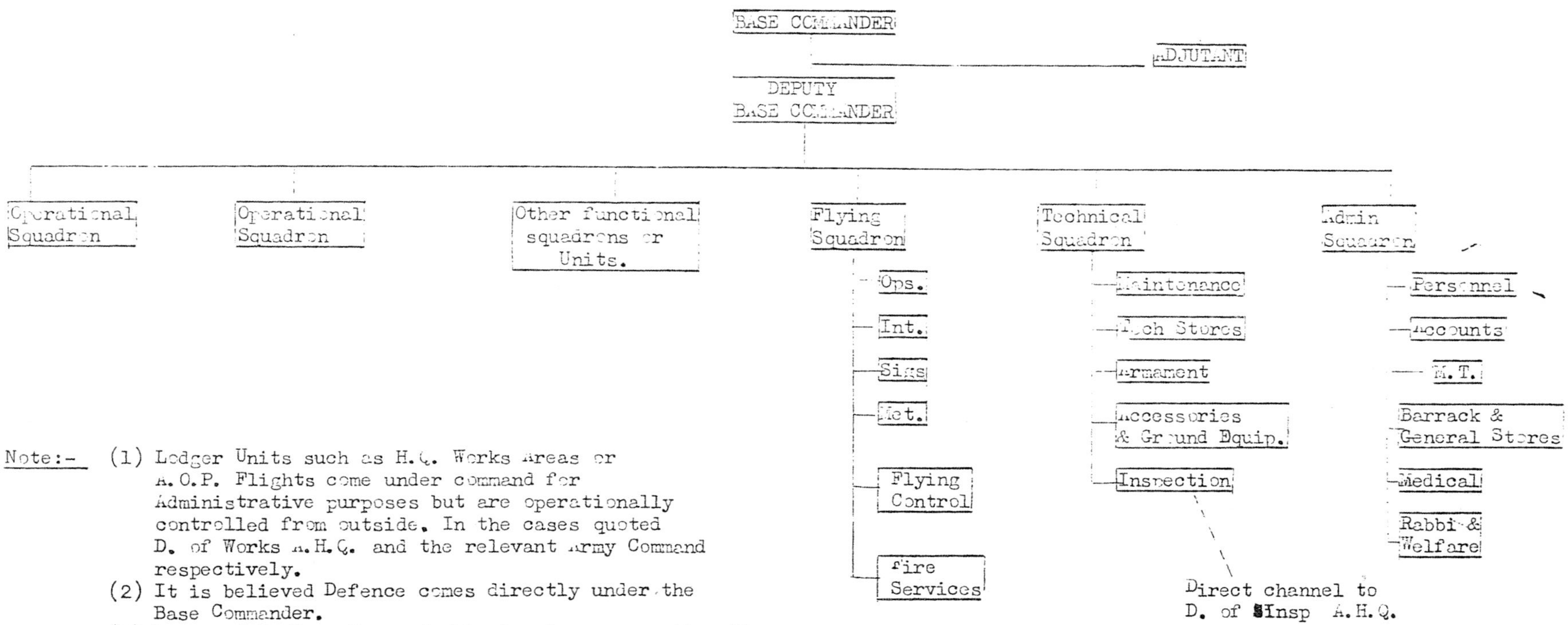

Note:-

(1) Ledger Units such as H.Q. Works Areas or A.O.P. Flights come under command for Administrative purposes but are operationally controlled from outside. In the cases quoted D. of Works A.H.Q. and the relevant Army Command respectively.

(2) It is believed Defence comes directly under the Base Commander.

(3) Though O.C. Base Commands his Squadrons operationally they are fought in the air from the Main Air Force Ops. room at A.H.Q. in conjunction with the radar controllers.

S E C R E T.

Appendix "B" to
AA.Tel/S.83/Int. dated
1st. December, 1957.

ISRAEL AIR FORCE - ORDER OF BATTLE

30th. NOVEMBER, 1957.

1. Ramat David - No. 1 Wing.

(a) Front Line Operational Units

Unit	Role	Aircraft		
109 Squadron	DF/GA.	Mystere IV A		24
117 Squadron	Peace - Jet	Meteor Mk. VII	5	
	Conversion.	Meteor Mk. VIII	13	
	War - DF/GA.	Meteor Mk. IX	3	21
119 Squadron	Night Fighter	Meteor Mk. XIII		3
		Total aircraft		48

(b) Non-Operational Units

Unit	Aircraft		
Station Flight	Mustang	2	(Drogue towing).
	Piper Cub	2	
	Stearman	1	
	Harvard	1	
Works Area H.Q. (Northern Area) -			
	Total aircraft	6	

2. Hatsor - No. 4 Wing

(a) Front Line Operational Units

Unit	Role	Aircraft	
101 Squadron	DF/GA	Mystere IVA	24
113 Squadron	Peace - Operational Training Unit. War - DF/GA.	Ouragan	24
		Total aircraft	48

(b) Non Operational Units

Unit	Aircraft		
Station Flight	Mustangs	2	(Drogue towing)
	Stearman	1	
	Harvard	1	
	Piper Cub	2	
	Total aircraft	6	

3. Ekron - No. 12 Wing

(a) Front Line Operational Units

Unit	Role	Aircraft		
103 Squadron	Transport	Dakota	15	
		Noratlas	3	
		Sikorski S55	2	20
115 Squadron	Photo Recce	Mosquito Mk XVI	8	
		Mosquito Mk III	2	10
		Total aircraft		30

(b) Reserve Operational Units

Unit	Role	Aircraft	
(@) Squadron	G/A	Mustangs	18
		Total aircraft	18

(@) There is now only one reserve Mustang Squadron.
This was previously No. 107 Squadron but is believed

SECRET.

-2-

Appendix "B" to AA.TEL/S.83/Int dated 1st. December, 1957.

3. Ekron - No. 12 Wing (continued).

(c) Non-Operational Units

Flying Training School	Mustangs	7	
	Harvards	24	
	Stearman	24	
	Oxfords	3	58
Central Maintenance Unit			
Air Force Staff School (or College)			
Air Force O.C.T.U. & School of Administration			
Central Medical Board			
Works Area Headquarters (Southern Area)			
P.T.C.			
Station Flight	Stearman	1	
	Piper Cub	2	3
	Total aircraft		61

4. Ramle

(a) Operational Units.

100 Squadron A.O.P. Piper Cubs

Ramle Squadron H.Q.		
Operational Flight A	4	
Training Flight	4	
Technical Flight	6	
A.H.Q. Comm. Flight	3	
Ramat David Operational Flight B	4	
Beersheba Operational Flight C	4	
Eilat Operational Flight D	4	
Tel Aviv Air Gadna Training	6	35
Total aircraft		35

(b) Other Units.

Headquarters Israel Air Force
Air Headquarters Unit
Air Force Intelligence Unit
I.A.F. Operational Control and Reporting Unit.

5. Haifa - No.21 Wing

Israel Air Force Technical Training School

6. Miscellaneous Units

Equipment Depot - Sarafand Camp
Radar and Radio Operators School - Sarafand Camp

/7.

S E C R E T.

- 3 -

Appendix "B" to
AA. TEL/S.83/Int dated
1st. December, 1957.

7. Reserve of Aircraft

Aircraft available as reserve to operational Units and training machine and not included in paragraphs 1 - 6 above, either in the hands of the Central Maintenance Unit, Bedek or in reserve on main bases.

Mystere IV A	(Available to 101 and 109 Sqdns)	7	
Ouragan	(for 113 Squadron)	4	11
Mustang	(Available for Reserve Sqdn. & F.T.S.)	5	
Harvard	(Available to F.T.S. & Station Flights)	12	
Stearman	(Available to F.T.S. & Comm. Flights)	15	32
	Total aircraft		43

8. Aircraft Aggregates

Total front line Operational Force	161 aircraft
Total Reserve Operational Force	18 aircraft
Total Reserve Aircraft	43 aircraft
Total Non-Operational	73 aircraft
Grand Total	295

Note :-

Under paras 1(a), 2(a) and 8 the Ouragans and Meteors are shown as Front Line Force despite their peace time training roles as they can be ready for battle in not more than twelve hours.

SECRET.

Appendix "A" to
AA.TEL/S.83/Int dated
1st. December, 1957.

SUMMARY OF AIRCRAFT STRENGTH BY TYPES AS AT
30TH. NOVEMBER, 1957.

With the Air Force, in Units or immediate reserve.

SERIAL.	TYPE.	STRENGTH	REMARKS.
1.	Mystere IVA	55	Out of 60 delivered.
2.	Ouragan	28	Out of 30 delivered.
3.	Meteors	24	Mk. VII 5; Mk. VIII 13; Mk. IX 3; Mk. XIII 3.
4.	Mustangs	34	
5.	Mosquito	10	Ten in service represent the Photo. Reconnaissance Squadron.
6.	Dakota	15	
7.	Noratlas	3	
8.	Harvard	38	
9.	Stearman	42	
10.	Oxford	3.	
11.	Piper Cub	41	
12.	Sikorski S.55	2	
	Grand Total	295.	

Aircraft no longer in Service.

SERIAL	TYPE	STRENGTH	REMARKS
1.	B.17	3	
2.	Hiller H.36	2	
3.	Spitfire	13	Ten at Haifa, three at Bedek.
4.	Commando	2	Sold to an American firm.
5.	Fokker S.11	35	
6.	Rapide	4.	
7.	Norseman	2	
8.	Oxford	14	
9.	Catalina	2	
10.	Mosquito	41	18 at Bedek; 12 at Kfar Sirkin; and about 11 at Ekron.

Color Section

The most numerous "warplane" early in Sherut-Avir—later Chel Ha'avir—was the British Auster lightplanes, mostly taken from RAF stocks. (Author)

The deHavilland DH-89 Dragon Rapide served in small numbers with the Sherut-Avir and later the Chel Ha'Avir (Barak)

The Kaydett was a beloved aircraft, both by the instructors and pupils, which praised its aerobatic qualities and ease of handling. (Barak)

During many years, even after retirement, a single Spitfire was kept airworthy, painted all black with a red cheat line.(Barak)

A few Sikorsky S-55s were used mainly in the search and rescue missions. Note that this machine has performed six live rescue missions and helped bring to the world a new baby!.(Barak)

(Above and Following) This beautifully restored Spitfire Mk.9 is preserved in the Israeli Air Force Museum in the markings of 105 Squadron, this unit kept them in service until replacing them with P-51D Mustangs in the mid fifties. Note the rocket racks below the wings.(Author)

This preserved dual seat Meteor is one of the rare T.7/8s acquired in Belgium which had the T.7 airframe, plus the F.8 tail unit.(Amram)

The Meteor NF.13s were the most advanced Israeli Air Force all weather fighters until 119 Squadron re-equipped with Mirage IIICJs in the early sixties, note the two types of camouflage applied, the early one was in brown and blue, while the later had the green, tan and sand. After being phased out a few NF.13 continued to be flown for experimental purposes. (Author)

P-51D Mustang number 38 (ex-Swedish Air Force at the Air Force Museum). The Sinai Campaign stripes are wrong, but noteworthy is the insignia of 116 Squadron insignia. Actually, this squadron is equipped with A-4N Skyhawks, and shortly will supplant them with F-16s.(Author)

In all, during the Sinai Campaign, the T-6s flew 24 combat missions, launching 320 rockets, before returning to regular training duties, note campaign identification stripes.(Author)

Black painted Spitfire number 57 was used during the sixties by Chel Ha'avir's Commanding Officer, Ezer Weizman, as his personal mount. Now it rests in the Museum at Hatzerim.

The MiG-15 was the main adversary of Israeli fighters during the Sinai Campaig. Illustrated is a MiG-15UTI two-seat conversion trainer preserved at the Museum.(Amram)

Israel received three variants of the twin-jet Vatour, the attack IIA (foreground), the all weather interceptor (middle, fitted with a prolongued radome for electronic warfare roles), and the reconnaissance IIR (background). The three variants were replaced by the F-4E/RF-4E Phantom II from the late sixties, early seventies.(Author)

The Harvards were to be replaced in the Flying School by the Potez Air Fouga Magister jets, built under licence by Israel Aircraft Industry. This yellow example is maintained in flying condition at the IDFAF's Flying School at Hatzerim air base.(Author)

Mystère IVA number 60 seen at the Museum in 1989. When the first Mystères arrived in Israel in 1956, they initially replaced the Mustangs from 101 Squadron. (Amram)

The Douglas C-47 Dakota, albeit receiving less limelight than the combat types, but has been in service since the beguinning of the Israeli Air Force, amassing tens of thousands of flight hours. Illustrated is 4X-FNL code 04 at the Museum (Amram).

The Super Mystère SMB.2s served during all their operational life with 105 Squadron, and later in their career saw their Snecma Atar 101 engine supplanted with the Pratt & Whitney J-52, being rechristened as Sa'ar (Storm).

Israeli Air Force insignia.

Israeli Air Force Wings.

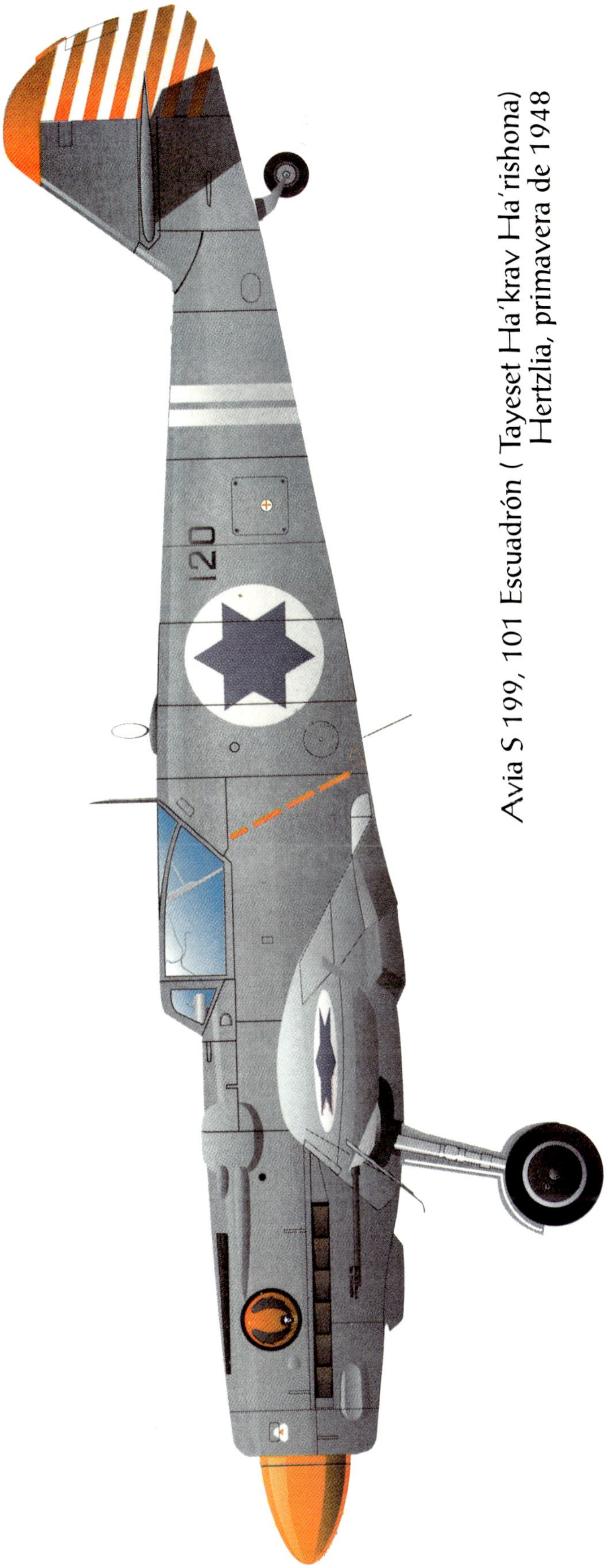

Avia S 199, 101 Escuadrón (Tayeset Ha'krav Ha'rishona)
Hertzlia, primavera de 1948

Avia S199, 101 Squadron (Tayeset Ha'krav Ha'rishona), Hertzlia airfield, Spring 1948. (J.A. Guerrero)

Supermarine Spitfire Mk IX (LF) 101 Escuadrón
Hertzlia, verano de 1948

Supermarine Spitfire Mk.IX(LF) 101 Squadron, Hertzlia airfield, fall 1948. (J.A. Guerrero).

Squadron Insignia

101 Squadron.

105 Squadron.

110 Squadron.

116 Squadron.

117 Squadron.

119 Squadron.

Also from the publisher